The Fight to Stay Put
Edited by Giorgio Hadi Curti, Jim Craine and Stuart C. Aitken

T0139765

MEDIA GEOGRAPHY AT MAINZ

Edited by
Anton Escher
Chris Lukinbeal
Stefan Zimmermann

Managing Editor
Christina Kerz

Volume 3

The Fight to Stay Put

Social Lessons through Media Imaginings
of Urban Transformation and Change

Edited by Giorgio Hadi Curti, Jim Craine
and Stuart C. Aitken

 Franz Steiner Verlag

Cover illustration: © Giorgio Hadi Curti

Bibliographische Information der Deutschen
Bibliothek
Die Deutsche Bibliothek verzeichnet diese
Publikation in der Deutschen Nationalbibliographie;
detaillierte bibliographische Daten sind im Internet
über <http://dnb.ddb.de> abrufbar.

ISBN 978-3-515-10212-4

© Franz Steiner Verlag, Stuttgart 2013
Gedruckt auf säurefreiem, alterungsbeständigem Papier.
Druck: Laupp und Göbel, Nehren
Printed in Germany

Geographers usually see themselves in a tradition of Alexander von Humboldt and we would like to highlight his position that everything in our world is interdependent. Deriving from that assumption geography has to be seen as an unlimited science, a discipline that knows its borders but has none.

Anton Escher, Chris Lukinbeal and Stefan Zimmermann,
Mainz, January 2008

CONTENTS

Jim Craine / Giorgio Hadi Curti

PROLOGUE: INTRODUCING THE FIGHT TO STAY PUT: SOCIAL LESSONS THROUGH MEDIA IMAGININGS OF URBAN TRANSFORMATION AND CHANGE

I work for a man named Lenny Cole
And Lenny Cole has the keys to the back door of this booming city
Let me give you an example of how Lenny works his magic
Two years ago, this property cost 1 million pounds
Today, it costs 5 million
How did this happen
Attractive tax opportunities for foreign investment…
Restrictive building consent and massive hedge-fund bonuses
London, my good man, is fast becoming the financial and cultural capital of the world
London is on the rise
Property value has gone one way: up
And this has left the natives struggling to keep a foothold in the property ladder
I can't teach you how to skin a cat
But I can tell you a lot about the money in bricks and mortar
Like he said, it's going one way

The overarching theme of our book is very much apropos to the above voice-over by Archy (Mark Strong) opening the Guy Ritchie film, *RocknRolla* (2008). How media portrays the struggle of "natives" – with an overt recognition that this term is always heavily conflicted and politicized – "to keep a foothold in the property ladder" is a central concern of this collection, as is the role media plays as an effective and provocative force of resistance and resilience in the fight to stay put. As may be obvious to many, the title of this collection – "The Fight to Stay Put" – is both an homage to and a play on Chester HARTMAN's 1984 manifesto, "The Right to Stay Put." It is here where HARTMAN first makes the case that property 'ownership' must be approached as much more than simply a purely economic matter. How we affectively become connected to place – and place's most intensive zone of return: the home – indelibly matters. HARTMAN (1984, 308) explains:

> Property someone lives in but does not own engenders a special relationship between user and property – destroying bonds built up through usage (often long-term usage) – produces large individual and social costs.

The destructive social costs and constraints imposed on people may be economic in nature, they may be political, or they can be viewed simply as an interruption of a social construct – whatever they are, they are always emotional and affective and, more often than not, indelible from historical and geographical legacies of privilege and marginalization.

Thus, on the one hand, fights to stay put can be seen as representations of struggles for social justice. On the other, they can be seen as the inevitable non-representational outcomes of territorial conflicts where fights for and over affective connections to place clash with those seeking monetary profit and power through re-productions of space and place. In a geographical sense, this collection considers how disadvantages suffered by certain sections of society relate to location. HARVEY (1973, 1996) characterizes such struggles as an attempt on the part of an oppressed or disempowered social group to play a viable role in the politics of spatial re-production and place construction. Standing formations of space and place provide some comfort of stability, a vibrating semblance of permanency that is always contingent upon social and spatial processes. Varying forms of institutionalized power, material practices and social relations all influence the degree of investment in a place, and when a group is forcibly distanced, displaced or unable to have a voice in spatial re-productions and place construction, conflict – the fight to stay put – intensively *and* extensively occurs (WILSON 1995).

Resulting fights highlighted in this collection point to how political, social and economic rights come to be and how necessary 'fights' for rights are not only captured, but work and are worked through various forms of media. In a world that is a continual battleground of social, political and economic injustices, the fights to stay put included here are some of the tools the disenfranchised or displaced have used in struggles against spatial misappropriation and redistribution. As forces of neo-liberal globalization continue to decentralize space and resultant redistribution of capital perpetuates various forms of inequality, struggles over property for the benefit of the few with the marginalization of the many will continue to occur. The trajectories of these struggles and how these struggles work through different concrete and imagined media landscapes form the basis of our chapters.

We are aware that approaching weighty questions such as urban transformation and change, legacies of privilege and marginalization, and impacts of gentrification and displacement may raise concerns over the validity of analyzing these very material issues through media (re)presentations. Perhaps reflecting such concerns, inside the front dust jacket flap of the English version of Taiyo MATSUMOTO's *Tekkonkinkreet* (2007) *manga*, a brief explanatory message greets the reader: "'TEKKONKINKREET' is a play on Japanese words meaning 'a concrete structure with an iron frame,' and it suggests the opposing images of concrete cities against the strength of the imagination." What are we to make of this message hinting at an "opposition" between the imagined and concrete life of cities? Does its language of iron frames and concrete structures point to an obfuscating and illusory nature of the powers of abstract imaginings? Perhaps it is a reflection of MARX and ENGELS' (1986, 47) critique of the transcendental nature of thought in *The German Ideology*, wherein they press for the necessity of analyzing "real-life processes" to understand the development and nature of the "ideological reflexes and echoes" that are imaginations and narratives.[1] Or maybe, like the massive body of urban literature influenced by MARX and ENGELS, it is suggesting that we must always work first and foremost with and through processes and structures of the "real-life" (i.e. grounded, material, concrete) city to impel or secure any hope of a real or potent politics of justice and emancipation.

Does media take us away from this possibility? In other words, is media simply another means through which we are drawn into ourselves, into our own myopic interests and withdrawn from the social world around us – in the process forsaking rights of community by arresting action and relation for matters of pure contemplation and isolation? Is media a tool of capitalist production through which its limiting and limited interests are perpetuated to separate powers and reinforce hegemonic ideologies of the existing order of things? In other words, are media objects merely productions of capitalist capture effecting a new religion of passivity – the latest and greatest opium for the masses? Does media remove us through imagined and narrativized urban fabulations from real material conflicts and concrete processes of alienation in the city? Is this collection – by realizing its life *through* media – dependent on a spectacular fetishism masking critical thinking and obfuscating true emancipation in, of and for the city? Is all of this, in the end, just a different form of commodity fetishism tenuously veiled in a not-so-thin sheen of academic jargon and ivory-tower pretentiousness which blinds us to the real-life processes and struggles of humankind? Is it a symptom of a life that has merely become "an immense accumulation of *spectacles*" where all that once that "was directly lived has receded into a representation" (DEBORD 1983, 7)?

Perhaps it all can be taken this way, but only if one chooses to ignore BENJAMIN's (BUCK-MORSS 1991, 265) position that the children in all of us have the creative power to engage with capitalist objects of production in different and differentiating ways; DELEUZE (1988, 1990) and NEGRI's (1991, 2004) liberating Spinozism which embraces the imagination as an immanent, political and embodied force inseparable from any constitution of human life; and MASSUMI's (1992, 101) argument that "[e]ven if a body becomes in the privacy of its own home, with no one else around…it is still committing a social act." Indeed, as Giorgio Hadi Curti explains in the opening chapter of this collection, *Tekkonkinkreet's* message of "opposition" is, in fact, far from a depreciation of the imagination. Rather, it is a reference to the principles of *yin* and *yang*[2]: a system of movement and mutual reciprocation – never domination – flowing through the parallel terrains of thought and action always (re)producing the concrete city.

---

1   The entire passage reads: "In direct contrast from the German philosophy which descends from heaven to earth we descend from earth to heaven. That is to say, we do not set out from what men say, imagine, conceive, nor from men as narrated, thought of, imagined, conceived in order to arrive at men in the flesh. We set out from real, active men, and on the basis of their real-life processes we demonstrate the development of the ideological reflexes and echoes of this life-process" (MARX and ENGELS 1986, 47).

2   In Japanese: in and yo.

Underscoring the material significance of imaginings to the life of cities, Andreas HUYSSEN (2008, 3) argues that urban imaginaries are not simply dismissible or distancing "figments of the imagination." Rather, they are "embodied material fact[s]…part of any city's reality." He relates that "[w]hat we think about a city and how we perceive it informs the ways we act in it." Similarly, VIGAR et al. (2005, 1392) tell us, "[t]he city has always been an 'imagined environment'…Ways of seeing cities have long been critical in shaping the form, experience and governance of urbanity." As all of this suggests, the imagination is a vitalistic and constitutive force influencing perception, discourse, thought and action. As such, it is indelibly linked to political negotiations and social transformations of city life. Yet, we must take pause here: to consider the imagination as a constitutive force of life apart from the body threatens the possibility of falling into the all-too-common dualist trap of separating and privileging mind from body and thought from action. Thus, the question of how to approach the implications of the imagination in the concrete politics of urban life becomes a metaphysical one.

Antonio NEGRI (2004, 13) explains that "in the history of western thought and particularly in the history of the bourgeoisie, metaphysics and politics are constructed together" and "all philosophies that fail to consider the fabric of human passions as the sole effective reality upon which political analysis can operate" (NEGRI 2004, 14) neglect the constitution and collective life of real, concrete power. What NEGRI is getting to here is that it is a metaphysics that takes into account the immanent and embodied materiality and power of the passions – including their indelible connections to imagination, thought and action – that "is the terrain on which the history of political thought must work" (NEGRI 2004, 13). In agreement with this, it is our position – and this collection's most intensive thrust – that different media experiences are central to both (re)formations and revolutions of exclusionary processes of urban life and to their perpetuation, because not only do they provide imaginative glimpses and insights into how passions and their affections conflict and relate, but, more vitally, because they produce passions and imaginations which can affectively (be)come to connect or disconnect (in) us.

To this end, Colin Gardner explores the creation of a specific collective utterance through a close reading of ALEA's seminal and paradigmatic film *Memories of Underdevelopment* (1968), itself adapted from Edmundo Desnoes 1962 novel "Memorias del Subdesarrollo." As Michael CHANAN (1990, 3), explains:

> Underdevelopment is an economic concept, referring to the relationship between a country with the status of an economic colony and the metropolis that colonizes it. The title therefore claims that now colonization is over. Underdevelopment has been replaced by the Revolution. But it turns out that the new title is somewhat ironic, for what the film shows is the way that people continue to carry the mentality of underdevelopment within them, how it weighs them down, and how it becomes a problem.

The film's main character, Sergio, exercises through his own fight his singular right to "stay put," and thus ends up as a case study in geographical limbo but also as the catalyst for a new form of cinematic 'ethics.' For Gardner, ALEA's film is structured dialectically via two parallel trajectories that both inflect and ultimately *infect* each other as we are forced to negotiate the creation of a new form of de-territorialized text and urbanity that defies easy resolution.

Following the theme of urban transformation and change further, Stuart C. Aitken discusses Maryhill, a traditional working-class neighborhood on the northwest side of Glasgow, where filmmaker Lynne Ramsay was born and spent her childhood. The neighborhood developed during the industrial revolution along a corridor that focused on Maryhill Road and included the Forth and Clyde Canal. In the 1970s, the neighborhood became a Comprehensive Development Area (CDA) to help alleviate housing pressures in inner city Glasgow slums. Ramsay's *Ratcatcher* (1998) documents a summer in Maryhill during the 1974 dustbin workers' strike. The young protagonist in the film deals with violence and death while fantasizing about relocating to a more peripheral housing estate and a house with "a bathroom, a toilet, an' a field." Aitken uses *Ratcatcher* as a foil to under-stand larger issues of permanency and desire, and the ways these contexts are thwarted by grand, over-coded and under-designed urban spaces like CDAs.

Building on the themes of violence and desire, Fernando J. Bosco looks closely at the plight of the 'disappeared' during the military dictatorships of Argentina. *Crónica de Una Fuga/Buenos Aires 1977* (*Chronicle of an Escape*) is a recent example of a film genre that seeks to document the abuses committed by the military government of Argentina between 1976 and 1983. The film recon-structs, through a script in part based on a memoir and historical research and in part fictional, the story of a group of young men who are abducted, imprisoned, and physically and psychologically tortured in an illegal detention center in sub-urban Buenos Aires during several months in 1977. It is an intimate film in which the emotional geographies of four men – and the socio-spatial relations they de-velop in the small confines of the prison/home they are forced to share – speak to larger issues of urban violence and resistance that were not only characteristic of Buenos Aires and Argentina during those turbulent years, but that speak in many ways to current urban conditions the world over.

Pascale Joassart-Marcelli expands on themes of emotion, desire, violence and permanency in her exploration of the ways in which undocumented immigrants fight to stay put – that is, how they negotiate and resist the multiple boundaries that preclude belonging and citizenship – through Jean-Pierre and Luc Dardenne's *La Promesse* (1996) and Tom McCarthy's *The Visitor* (2007). In the face of intolerance and cruelty, both films acknowledge the multiple spaces and openings where belonging can occur and illustrate the role of emotions and social interactions in defining these spaces. We find that citizenship can consist of multiple layers (ethnicity, religion, race, gender, etc.) and intersecting scales (home, neighborhood, city, nation, global communities), and Joassart-Marcelli uses the two films to show a different immigrant politics that moves the struggle beyond traditional rights and duties toward what Young calls "differentiated citizenship" (1999, 264), which acknowledges and respects the different positions, interests and identities of citizens. The fight to stay put thus demands a new politics of citizenship that focuses on creating and preserving places of belonging that affirm difference.

Turning our attention to representations of the city, Amy Siciliano and Paul S. B. Jackson explore the role of *Sesame Street* as an American cultural response to urban crises of the 1960s. Specifically, the authors consider how the urban fantasyland projected through this children's educational television program helped reshape meanings associated with the city and urban social identities. The authors suggest that *Sesame Street's* sanitized 'ghetto' aesthetic and sterilized social scripts helped to generate a new image of the city as a desirable locale for the white middle class. In projecting an idealized New York streetscape into the homes of millions of Americans, it helped alter public discourses and representations about urban space and middle-class identification with these spaces.

A companion piece is Steven M. Graves' discussion of gentrified televisual landscapes. Several widely popular TV programs airing during the 1970s helped bolster the potential ground rent of urban housing stock by altering the symbolic value of specific housing types. Demand for late 19th and early 20th century Victorians, Brownstones and the like was enhanced by TV programs that housed highly charismatic television characters, who were often young, chic, professional and childless. Simultaneously, television producers frequently placed TV families from other points in their life cycle or career arc in landscapes and housing stock that were clearly not gentrified. Thus, as Graves indicates, television programming has been a significant factor in the generation of demand for specific types of housing stock, and perhaps the diminution of demand for other types of housing. For the masses accustomed to viewing American landscapes from the comfort of their homes, television made gentrified living appear highly desirable without ever exposing the many hidden costs associated with the gentrification process.

Expanding on these themes of media, its material connections to urban transformation and change and the fight to stay put, Rachel Goffe and Todd Wolfson look at the unfolding practices of an organization based in Philadelphia to explore how building a media and communications infrastructure can begin to collapse the material and discursive distances between people while connecting the struggles they face in their everyday lives. The Media Mobilizing Project (MMP) engages poor and working people in telling the untold stories of their lives to each other through web, radio, video and TV platforms, as well as through face-to-face meetings and media-making; building an infrastructure of technology and relationships between people and groups; while also creating access to media, technology and the means to distribute information and stories. The fight to stay put here is broadly conceived by Goffe and Wolfson as the rejection of a future for Philadelphia shaped by the needs of capital rather than those of people. The statement, "Movements begin with the telling of untold stories," is used by MMP in describing a vision for media's role in movement building. The day-to-day sharing of information between organizations and individuals thus forms the basis for lasting relationships, a shared leadership and a shared strategy, out of which grounded and concrete institutions of representation can emerge – institutions that can help upset capitalism's subordination of life to its role as the source of value.

The next chapter further explores relationships between media technology and social movements and connections. Maude Ontario is author Amanda Huron's DJ name. She is a founder and active member of Radio CPR and in her chapter she discusses the use of microradio as a weapon in the fight to stay put. As Huron states, a radio wave appears to be fleeting – it cannot be seen or touched, apparently ungrounded, an ethereal presence detached from the earth. Yet radio in its smallest forms can be deeply connected to the land. Huron examines the case of one contemporary microradio station in its struggles against neighborhood displacement and finds that the particular geography of microradio can be a powerful tool for fighting for the right to be in a certain place: the right to stay put over time, to create culture, to dwell.

Bringing together themes of city representation, social movements, emotion and passion, Ryan J. Goode, Kate Swanson and Stuart C. Aitken engage neo-Marxist theorizing from Benjamin and Williams to Deleuze and Lefebvre to trace media culpability in the continuing transformations of Rio de Janiero's favelas. Using *Cidade de Deus* (*City of God*; 2002) as a starting point, the authors outline a corporate global media blitzkrieg that flows through Rios' successful attempt to attract the 2016 Olympics to a point where institutional authorities are able to place Pacification Police within *favelas*. With a specific focus on the *favela* of Santa Marta, the authors examine the ways pacification is about occupying streets and reigning in youthful passions, while also pointing to how young residents fight back through the media components of *Visão Favela* – a grassroots organization that moves beyond Santa Marta to embrace a larger swathe of Rio's social and spatial *favela* life.

Michaela Benson and Ed Jackiewicz explore the prospects of 'the right to stay put' as a concept for explaining the impact(s) of lifestyle migration and residential tourism development, a matter indelibly entangled in urban representations presented through and promoted over internet websites. With an empirical focus on migration and development in Panama, the authors question whether 'the right to stay put' can adequately account for the various impacts felt on a local level. Benson and Jackiewicz find some evidence of displacement coinciding with the recent in-migration of North American and European populations, but understanding this phenomenon within the wider context of social and economic change in Panama reveals that the reality is far more complex than 'the right to stay put' paradigm allows.

Continuing with the theme of media representation in a different way, Brenda Kayzar suggests that representations in the media depicting philanthropic work as having saved neighborhoods overlooks the continued injustice in planning practice and the efforts of community members to agitate for some reasonable level of justice in the provision of municipal services and maintenance. Using the City Heights neighborhood in San Diego, California, Kayzar juxtaposes the ways in which the media represented the community and reinvestment outcomes once a philanthropic organization was enrolled in improvement efforts. Kayzar demonstrates that amidst a long struggle for justice, community members overcame language and cultural differences and banded together in activism. They established a community development corporation (CDC) that informed the city of the community's needs through small-scale efforts such as organized clean-up, neighborhood watch programs, a community garden, and an annual street festival. The CDC was also the portal through which the community strongly expressed their needs for a police station and schools, and these needs were represented in a community plan and during 'retreats' with civic leaders. All of this represents the 'groundwork' laid for ongoing changes in the community. The groundwork demonstrates that the citizens of City Heights were not simply a community out of control, as was suggested in media accounts that focused on crime and poverty. Citizens were active and had expectations and they expected equitable treatment in planning practice and service provision. Yet from the privileged site of the media, the accepted discourse credits the philanthropic organization for procuring necessary city resources, thereby becoming the activist and savior of the community.

Jason Dittmer brings the world of comic books into our discussions of urban representations, race, space and violence. Using a later version of American comic book hero Captain America, he addresses the ways in which race inflects both representations of displacement and gentrification in American popular culture and also the ways in which a 'right to stay put' might be implemented in the United States. In so doing, Dittmer highlights the potential for inequality should such a right be theorized and implemented strictly on the basis of neo-Marxist understandings of class.

Weaving together several themes of this volume, we consider in our final chapter how 'monstrous' bodies and landscapes in David Cronenberg's film *Videodrome* are created and how desires, subversions and violence are inscribed in bodies and the city – metaphorically and literally. The main question moving us through this chapter is: how are *we* created in and through our connections to media? The fight to stay put is different in *Videodrome* insofar that we must not only reinterpret new realities because they are constructive processes, but must fight with the production of our own selves to materially perform difference through activity and creativity. In other words, this fight is about the greater need to confront, encourage, and/or challenge different technological and scientific realities and their capacities to impel new geographies of monstrous effect and desire. Doing so provides us with a double understanding of how the science and technologies of media work as creative, but also limiting, manipulative and hegemonic, forms effecting economic, spatial and social influence in (if not control over) the changes and transformations of our cities and our world(s).

By approaching urban transformation and change through media's imaginative capacities to bring about different social lessons and material connections of and to urban life, we wish for this collection in both its component parts and as a collective whole to add to critical perspectives on the life of cities through what Brian MASSUMI (2002, 12–13) calls "a productivist approach," or affirmative methods and techniques "which embrace their own inventiveness and are not afraid to own up to the fact that they add (if so meagerly) to reality." Part of this inventiveness is to not simply be content with established codings of concepts (e.g. "the right to the city"; "the right to stay put"; "gentrification and displacement"; "class", etc.) but, instead, to "uproot" them and experiment with the "residue of activity from [their] former role[s]" (MASSUMI 2002, 20) by "reconnect[ing them] to other concepts, drawn from other systems, until a whole new system of connection starts to form" (MASSUMI 2002, 19).

To be productivist is to be creative; and to be creative is to be critical. However, as MASSUMI (2002, 12) warns, if it is difference we desire to make, that is, if our move is towards a material and spiritual human emancipation of greater relational capacities between people and people and people and things, the "critical thinking" of convention and tradition is of little value as it "disavows its own inventiveness as much as possible" and "it carries on as if it mirrored something outside of itself with which it had no complicity." In this sense, conventional critical thinking does not and can not be a move of or towards human emancipation because it re-produces, and in the process rationalizes and validates, the limiting bonds of the very material and capacitational boundaries it seeks to subvert.

To be critical, then, cannot be to work through closed and pre-established codes, transcendent rules of action or in the interests of already-constituted groupings. Rather, it must be to confront the poverty of both political and creative imaginations and joyfully become through the emergence of differentiating relations between people and people and people and things: a cultivation of good bodily affections and affects through the milieu of the social as "more than human" (WHATMORE 2002). As Spinoza tells us, it is such cultivations of joy that allow us to think, feel, accomplish, know, create, *do* more in and as a community of a multitude; and as DELEUZE and GUATTARI (1994, 48) tell us, it was Spinoza who first "discovered that freedom exists only within immanence." Thus, to be critical – and for media to work as a critical machine materially opening up new possibilities and potentialities of emancipation – we must not engage with or approach media merely as matters of description, representation, production or ideological critique, but, to productively and contextually borrow from Warren MONTAG (1999, 21), as "a body among other bodies, and, if it is effective, 'moves' other bodies to act or to refrain from action, or, more properly, moves bodies and minds…to act and think simultaneously."

REFERENCES

BUCK-MORSS, S. (1991): The Dialectics of Seeing: Walter Benjamin and the Arcades Project. Cambridge.

CHANAN, M. (1990b): Tomás Gutiérrez Alea: A Biographical Sketch. *Memories of Underdevelopment, Rutgers Films in Print Series, Vol. 15.* New Brunswick & London, 15–22.

DEBORD, G. (1983): *Society of the Spectacle.* London.

DELEUZE, G. (1988): *Spinoza: Practical Philosophy.* San Francisco.

DELEUZE, G. (1990): *Expressionism in Philosophy: Spinoza.* New York.

DELEUZE, G. AND F.GUATTARI (1994): *What is Philosophy?* New York.

HARTMAN, C. (1984): The Right to Stay Put. GEISLER, C. C. and F. J. POPPER (Eds.): *Land Reform, American Style.* Totowa, 302–318.

HARVEY, D. (1973): *Social Justice and the City.* Cambridge.

HARVEY, D. (1996): *Justice, Nature and the Geography of Difference.* Cambridge.

HUYSSEN, A. (2008): Introduction. HUYSSEN, A. (Ed.): *Other Cities, Other Worlds: Urban Imaginaries in a Globalizing Age.* Durham.

MARX, K. and F. ENGELS (1986*): The German Ideology,* ARTHUR, C. J. (Ed.).New York.

MASSUMI, B. (1992): *A user's guide to capitalism and schizophrenia: deviations from Deleuze and Guattari.* Cambridge.

MASSUMI, B. (2002*): Parables for the virtual: Movement, Affect, Sensation.* Durham.

MATSUMOTO, T. (2007): *Tekkonkinkreet: Black & White.* San Francisco.

MONTAG, W. (1999): *Bodies, Masses, Power: Spinoza and his Contemporaries.* London.

NEGRI, A. (1991*): The Savage Anomaly: The Power of Spinoza's Metaphysics and Politics.* Minneapolis.

NEGRI, A. (2004*): Subversive Spinoza: (Un)Contemporary Variations.* Manchester.

VIGAR, G., S. GRAHAM and P. HEALEY (2005): In Search of the City in Spatial Strategies: Past Legacies, Future Imaginings. *Urban Studies* 42 (8), 1391–1410.

WHATMORE, S. (2002): *Hybrid Geographies.* London.

WILSON, D. (1992): Urban Conflict Politics and the Materialist-Poststructuralist Gaze. *Urban Geography* 16 8*, 734–742.

Giorgio Hadi Curti

"THIS IS MY TOWN" – EXPLORING THE AFFECTIVE LIFE OF URBAN TRANSFORMATION AND CHANGE VIA TAIYO MATSUMOTO'S AND MICHAEL ARIAS' *TEKKONKINKREET*

This is my town.
~ Black

This is my town!
~ Kimura

My town must be transformed.
~ Snake

Watch what you say with the "my town" talk.
It's a bad habit…This isn't anybody's town.
~ Gramps

"Whose city is it?" – This is a question, Saskia SASSEN (1996, 206) explains, that emerges out of a double opening: on the one hand, for new capacities of relation and connection across and over space; on the other, for an increase in conflicts and negotiations of and over place. Taiyo MATSUMOTO's *manga* (1993–1994; English language version 2007) and Michael ARIAS' eponymous animated filmic adaptation (2006 [2007]), *Tekkonkinkreet* (*Tekkon kinkurîto*), together provide an imaginative and powerful glimpse into the ways such connections and disconnections, conflicts and negotiations open up and (un)fold through the movements and moments of urban transformation and change today taking place (and taking-place) in cities the world over.[1] Like SASSEN, *Tekkonkinkreet* speaks of the city through the language of ownership; that is, through the social relations and nego-tiations of power and control defining who has the 'right' to call and make the city their own. It also speaks through intensities of *affection and affect*; that is, how the city horizontally emerges as a concrete practice by way of intensive and extensive capacities and relations between bodies and spaces, desires and ideas. It is especially through its subterranean and unrestrained powers of affection and af-fect – matters which have curiously received little explicit attention in the urban literature to date (THRIFT 2007, 171; ANDERSON and HOLDEN 2008) – that *Tekkonkinkreet* hints at a different sensibility that might be taken to questions of urban transformation and change. It is partly for this reason that I open up this volume on media imaginings of *The Fight to Stay Put* with *Tekkonkinkreet*. But, perhaps more vitally, it is because *Tekkonkinkreet* is the seminal fluid that im-pregnated and propelled this project from a singularized, almost imperceptible shadow of an idea into an illuminated shared conceptual framework and collective material flow.

DEVELOP OR PERISH, GENTLEMEN!

The life and times of *Tekkonkinkreet* (un)fold in and through Treasure Town, the third district of an unnamed Japanese city that is simultaneously an imagined urban singularity and a concrete global condition. It is an architecturally and fan-tastically syncretic town, drawing aesthetic and formal inspiration from a multi-plicity of cultural and religious milieus – from the Islamic calligraphy of South-west Asia to the divine iconography of Hindu India to the cityscapes of French graphic-novels to the streetscapes of contemporary Japan – it is a town that is both everywhere and nowhere at once; it is a town *yakuza* member Suzuki "the Rat" tells us has a certain *"je ne sais quoi."*

---

1   While the film version of *Tekkonkinkreet* is a close adaptation of the manga, there are slight differences in narrative structure and certain character developments/personalities. These differences, however, create little contradiction in or to the narrative. As such, I treat the manga and film versions as two parts of the same universe and draw on the dialogue and narratives of each. Where there is divergence or discrepancy, I privilege the filmic narrative. When language from the manga is inserted into dialogue from the film it is placed in parentheses.

At the core of this urban 'certain something', this singularly affective 'I do not know what', is a brewing conflict over ownership: a global syndrome facing cities the world over that manifests itself in unique ways in and through the distinct particularities and everyday processes of different social and spatial milieus; a *syndrome* being "the meeting-place or crossing-point of manifestations issuing from very different origins and arising within variable contexts" (DELEUZE 1989, 14), and a *milieu* being a surrounding, a medium and a middle (DELEUZE and GUATTARI 1987). Time is told in *Tekkonkinkreet* through the movement of the seasons: from summer to fall to winter to spring to summer again – a circular return of difference confronting the city that we have seen before (though never exactly in the same way). Corporate interests are moving in (from who knows where) and taking over (we *do* know what – *space*). These interests, in partnership with local yakuza leader, the Boss, and his henchmen, the Rat and Kimura, are led by the manipulative and sanctimonious Professor Snake – a character simultaneously embodying the impersonal coldness and conceited ethic of global neoliberal capital, the "god-trick" (HARAWAY 1991) of the universalizing dogmas and modernist principles of urban planning and development, the insidious violence of 'creative destruction' and the self-righteous and exclusionary ideologies of economic religiosity (see NELSON 2001). Snake proselytizes to Kimura mid-way through the film:

> **Snake:** I'm going to change this city. And it makes no difference if you are in or out…Alliances with cops, yakuza, are important, but those alone won't change this town. My town must be transformed.
> **Kimura:** Your town?
> **Snake:** Few are chosen to pursue the ideal.
> **Kimura:** What are you after?
> **Snake:** We serve the ultimate authority. Do you know who that is, friend?...God.

This pronouncement of ownership by Snake is expressive of an urban planning from 'nowhere' and possession of everywhere, where reification of form and omniscient design stand in for real knowledge and affective connection, trumping any differences in, of and to the city. It is "the judgment of knowledge," DELEUZE (1997, 127) tells us, that transcends any *je ne sais quoi* of spatiality and "envelops an infinity of space, time, and experience that determines the existence of phenomena in space and time…[It] implies a prior moral and theological form, according to which a relation was established between existence and the infinite following an order of time: the existing being as having a debt to God." Today, this absoluteness of judgment and the Divine has become the totalistic morality of a neo-liberal "economic theology" (NELSON 2001) whose judgments are beholden to divinations of benefits and costs: a pathology wrapped and packaged as panacean remedy for the human and urban condition. The debt is now to the supposed 'objectivity' of science and the order of time is now the transcendent and fixed uni-linear progression-time of 'development' (ESTEVA 1992) whose only supposed alternative is death:

> **The Boss:** It's gratifying to see you all [different economic and political players in the city] here instead of busting each other's teeth on the pavement. Now is a time for new approaches. So, what does the world need most? The economy is stagnant. We've become white elephants. We need new business, new stimulation. Entertainment! This is the wave of the future! And what better place to start than with the piggy banks of kids and the wallets of their parents! Working with Mr. Snake has taught me a great deal, such a man of vision! A culture shock of sorts. But of the many gains we achieve the greatest is that we face our future. Develop or perish, gentlemen! And so, to continue…Professor Snake, if you would.
> **Snake:** Let's cut to the chase. The third district that you control, the so-called Treasure Town district…is…well, quaint at best. Unspoiled land, surrounded by the biggest growth areas in the city. That's where we come in. As you can see, since launching the first Kiddie Kastle project three years ago we've had nothing but unencumbered profit margins.
> **The Boss:** You truly have the Midas touch Monsieur Snake. And we thank you! A new revenue source! More wallets to pick clean!

Snake's plan to consume Treasure Town – or more exactly, to transform, gentrify and commodify it into a space of pure consumption (see CURTI et al. 2007) – is an attempt to arrest the spontaneous life of the city and subvert it to the order of structured and exclusionary development. Through complete demolition and the building in its place an expansion of *Kiddie Kastle* – an amusement park doubling as a monstrous parasite of a "symbolic economy" (ZUKIN 1995) out of control and a 'creative destruction' run amuck (or taken to their logical ends; the two, in the end, equating the same thing) – city space is to be transformed into an ever expanding horizon of economic profit. This plan will not, however, be carried out unopposed.

The Rat is repeatedly confronted by the police detectives Fujimura, a former judo champion and veteran cop from Treasure Town, and Sawada, a rookie with a "frigid" disposition from Tokyo who is on the lookout for some action. There are also Black and White (*Kuro* and *Shiro* in Japanese), the two members of "the Cats," or "Strays. Orphans. Delinquents who rule Treasure Town," Fujimura tells us, "They live by the law of the jungle…[and t]his city is their playground. Underestimate them and you'll kiss asphalt." While White hopes for a better future – "When this piggy [bank] gets fat we'll ride in an airplane. Right, Black? Build a house on the beach. That's the plan, right?"; "Be happy be happy" – and inevitably feels guilt over the consequences of their violent acts – "When we hurts people, I tell God we're sorry. I tell God we're sorry. Never do it again. Sorry. Sorry. Never do it again," Black lives in each moment and embraces his visceral strength – he is the "soul of [the] city"; for him, the fight for control of Treasure Town is "not a question of worrying or of hoping for the best, but of finding new weapons" (DELEUZE 1995, 178). It is only Black's love for White that tempers the potential of an absolute deterritorialized darkness brewing within him – "[Black]'s already lost faith in living. His only purpose in life has been to protect White. Without White, he'd have no reason to live." And in many ways it is such relationships of *Tekkonkinkreet* – between Black and White, between the Rat and Treasure Town, between the Cats and the yakuza, between Snake and Kimura – that *are* the city – the affective yin and yang of urban life, embodying both the hope and need of a loving and joyful *urbanitas* and the violence and fear of an urbanity imbued with exclusion, isolation, alienation and sadness.

AFFECT AND DESIRE: THE CITY

THRIFT (2007, 171) explains that "[c]ities can be seen as roiling maelstroms of affect. Particularly affects like anger, fear, happiness and joy are continually on the boil, rising here, subsiding there, and these affects continually manifest themselves in events which can take place either at a grand scale or simply as a part of continuing everyday life." Despite this ubiquity and the everyday intensive and extensive magnitudes of affect, little urban literature confronts the functions it plays in the life of cities (THRIFT 2007, 171; ANDERSON and HOLDEN 2008), a particularly curious oversight considering its central role in the politics and negotiations of place (see CURTI et al. 2007; CURTI 2008a; CRAINE and CURTI 2009). THRIFT points to three reasons for this overt lack of engagement: first, the residue of a "cultural Cartesianism" and its connected minimization of affect's importance; second, a "division of labour" between the "creative" arts and sciences;

and third, the difficulty of "capturing in print" qualities and intensities of affect. It is particularly THRIFT's first point – the residue of Cartesianism – which is in need of more direct attention,[2] because to remain in its residue is at best to bring affect back to the relatively impotent arena of representation, interpretation or meaning; and at worst, to continue to neglect its forceful role in negotiations of the city, and thus to miss out on the socially constitutive and fundamentally political and material natures of affections and affects (see AMIN and THRIFT 2002; THRIFT 2007; CURTI 2008a; ANDERSON and HOLDEN 2008).

Affect, unlike emotion, does not presuppose a subject or even a cultural realm (see MASSUMI 2002; THRIFT 2007). Rather, it transforms and displaces forces of memory and emotion constituting subjects (see CURTI 2008a) and is first and foremost a question of ethics. DELEUZE (1995, 100) explains that what distinguishes ethics from morality is that it is "the styles of life involved in everything that make us this or that," the horizontal relations of becoming(s), as opposed to "a set of constraining rules…that judge actions and intentions by considering them in relation to transcendent values." Ethics, then, is concerned with emergent capacities – of processes, for justice, for emancipation – while morality is concerned with eternal judgments – of behaviors, of actions, of exclusionary interests. In *On the Jewish Question* (MARX and ENGELS 1978), MARX distinguishes between two forms of emancipation: political and human. The first, a moral and localized move towards a political equality of rights; the second, an ethical and universal move towards freedom from the determining actions and limiting claims of capitalism. MARX explains that only the latter move is truly emancipatory, as it frees us from the separation of social and political power perpetuated by civil society and the political state (MARX and ENGELS 1978, 46). Indeed, in many ways MARX's entire project can be understood as working through an ethics critiquing "abstract bourgeois morality that divorces itself from the analysis of its own concrete historical structures of inequality and injustice" (MCCARTHY 1985, 197), while simultaneously working towards "the very conditions for the possibility of self-realization and actualization, for freedom (both individual and social), and for social justice" (MCCARTHY 1985, 182). In this way, ethics can be understood as an *ethology*, or the study of "compositions of relations or capacities *between* different things" (DELEUZE 1988, 126, stress mine). As ethology is always concerned with immanent forces and relations it is informed through capacities and physics of affect – affect always being material, embodied, pre- and post-subjective and whose existence always pre-supposes an *affection* – or, "the instantaneous effect of a body upon [another body]" (DELEUZE 1997, 139).

Affect thus moves beyond Cartesian notions of isolation and compartmentalization to speak through communities of bodies living together as the differential and emergent correlates constituting the *capacities* of place. In other words, affect is always a question of the production of difference by and through the emergent

---

2   Especially because it is a Cartesian ontology on which THRIFT'S latter two points ultimately
     depend.

relations and spatialities of our affecting each other and (be)coming together. In an explication of Spinozan thought, DELEUZE (1988a, 27) writes:

> [F]rom the viewpoint of an ethology of man, one needs first to distinguish between two sorts of affections: *actions*, which are explained by the nature of the affected individual, and which spring from the individual's essence; and *passions*, which are explained by something else, and which originate outside the individual. Hence the capacity for being affected is manifested as a *power of acting* insofar as it is assumed to be filled by active affections, but as a *power of being acted upon* insofar as it is filled by passions. For a given individual, i.e. for a given degree of power assumed to be constant within certain limits, the capacity for being affected itself remains constant within those limits, but the power of acting and the power of being acted upon vary greatly, in inverse ratio to one another.

Thus, there is a yin and yang relationship to affect, where activity – defined as an *action* – and passivity – defined as a *passion* – both in response to affections/effects of other bodies and in inverse relationship to one another – find their mutual impressions in what SPINOZA (Ethics, III, Prop. 9, Schol.) calls "appetite," or the consciousness of appetite, *desire*; which "is nothing but the very essence of man, from whose nature there necessarily follow those things that promote his preservation." As matters of "preservation," passion and desire are inseparable from the *conatus*, or the striving and endeavoring to persevere in being of all things (see DELEUZE 1990, 230–231). Because we live with others, because we are social beings, we necessarily affect others through our powers and capacities to affect, but so, too, are we necessarily affected by others through our capacity and power to be affected. Accordingly, SPINOZA (Ethics, III, Prop. 59, I) refines his definition of desire: "by the word *desire* I understand any of man's strivings, impulses, appetites, and volitions, which vary as the man's constitution varies, and which are not infrequently so opposed to one another that the man is pulled in different directions and knows not where to turn." Desire, then, is not an individual (though it may be a singular) force set against pre-formed or teleological social structures or beliefs, as it "is never an undifferentiated instinctual energy, but itself results from a highly developed, engineered setup rich in interactions" (DELEUZE and GUATTARI 1987, 215). In other words, desire is not superfluous to social, economic or spatial infrastructures. Rather, it is an indelible part of any given infrastructure itself (DELEUZE and GUATTARI 1983, 104). The question of desire then is never one of lack, but of affirmative capacity; the question of affections and affects thus becoming: what are they doing and for whom?

DELEUZE (1988a, 27–28) explains that a Spinozan sociality of passions/desires must be approached in two ways: as that affection/affect which brings *sadness* to a body – or that which diminishes or limits the capacity of a power to act; and as that affection/affect which brings a body *joy* – or that which is an increase or enhancement in a power to act. While the passion that is *joy* is passive, as it is still caused by an external affection, the "power of acting is nonetheless increased proportionately; we 'approach' the point of conversion, the point of transmutation that will establish our dominion, that will make us worthy of action, of active joys." Thus, affection and affects are always questions of how power increases and decreases in and through relations with others and the city is an

intensive milieu that can "be seen as a kind of force-field of passions that associate and pulse bodies in particular ways" (AMIN and THRIFT 2002, 84). In these
ways, affections, affects and passions/desires are immanently ethical and political
matters assessed in relation to what they do in helping and aiding a given body's
power and capacity to act, or by diminishing and displacing what a body can do
by subtracting or limiting its capacity or power; the problem then becoming a
question of "love and hate and not judgment" (DELEUZE 1997, 135) because, as
SPINOZA explains, "we neither strive for, nor will, neither want, nor desire anything because we judge it to be good; on the contrary, we judge something to be
good because we strive for it, will it, want it, and desire it" (Ethics III, Prop. 9,
Schol.). Questions of the city, then, are always questions of the ethical differences
of capacities and intensities of passion and desire ebbing and flowing in and
through communal encounters of relation and displacement; and it is through such
encounters of passion and desire in hope and fear, love and hate, joy and sadness
that the dual nature of affection/affect in and of Treasure Town (un)folds.

City Affection(s) / City Affect(s)

Fujimura knows the Rat, and more importantly, he knows what the Rat is capable
of doing: "He can change the entire personality of a city." The Rat's return is the
embodied manifestation and intensive ordinate of a global syndrome; his mere
presence a symptom and a sign: "a symptom being a specific sign of an illness"
(DELEUZE 1989, 14); a sign being "an effect"; an effect being "first of all the trace
of one body upon another, the state of a body insofar as it suffers the action of
another body, It is an *affectio*" (DELEUZE 1997, 138); and an *affectio* [*or* affection]
always being a displacement of ratios of joy and sadness. Fujimura's desire for
peace and order and affective connection to the city is displaced by the concerns
the Rat's affections raise in him. For Fujimura, concern congeals into apprehension when he and Sawada confront the Rat outside his car:

> **Fujimura:** I thought I smelled something sour in town.
> **The Rat:** Well, if it isn't my old friend, Fujimura.
> **Fujimura:** Nice little entourage, Rat.
> **The Rat:** This town's about to buzz like a bee hive poked with a stick. You'll regret being a
> cop now that I'm back.
> **Fujimura:** I already do.

Like Fujimura, Black is concerned with the Rat's return. He knows the Rat's
presence is a sign of violent social and urban change-to-come – it is his most intensive and pervasive effect; it is *what the Rat does*. As he sits high atop a perch
and surveys the city, he is the first to catch sight of the Rat, who, along with
Kimura and two other yakuza, are exiting their car; "The Rat's back," Black says
quizzically to himself. He soon sees the Rat a second time. At this moment he is
with the Boss at a kick-boxing match, and Black's face displays a heightened
sense of aggravation and disapproval; through durations of affect, his connection
to the city is becoming emotionally displaced. Outside, Black and White come

upon Kimura arguing with the Apaches, a local gang, and their leader, Choco: "We're the Apaches and we run Treasure Town! Choco's the name so don't forget it!" It is readily apparent that Black recognizes Kimura: "That yakuza…He's one of the Rat's men."

The yakuza are attempting to get the locals to deal their "stuff" (presumably drugs), but The Apaches refuse: "Try getting your hands dirty once in a while. We do all the dirty work! You listening!?" Choco shouts at Kimura. But Kimura's attention is briefly arrested by Black as the two of them lock in a momentary stare, their faces signaling the intensive durations of deterritorializing worlds; the deterritorializing forces: bodily affection and affect. Kimura's consciousness snaps back to his scheming attempts to wrest control from the Apaches. "Got no ears! Or balls!" Vanilla shouts. "Let's go," Kimura tells the other yakuza with a grin. "Hey, Kimura…We're not even gonna rough them up?...Suzuki won't like us coming home empty-handed" one of the yakuza matter-of-factly states. Kimura responds: "I'm running the show today."

As Kimura and the other yakuza drive away, Black asks: "Who was that Choco?"; Choco: "Some punk named Kimura." In the midst of the Apaches' cheers at their (temporary) victory, Black slowly stares in the direction of the departing car and says quietly to himself, "Kimura." As affect reterritorializes and solidifies into emotion, the affection which is its presupposition now has a name: "Kimura." So, too, has Black's presence made an indelible impression on Kimura: "Black…Suzuki said something about him. Some brat in Treasure Town (who can't tell right from wrong…). Likes the taste of blood. Pretty sure his name's…*Black*." Black and Kimura's paths will soon cross again.

YIN AND YANG: THE CITY

Though with different intent and purpose, what the social is for both the yakuza, Kimura, and the police detectives, Fujimura and Sawada, parallel one another, and oppose that of Black and White. As cops and yakuza, sociality is a code, a duty, a hierarchy, a structure – in short, social action is reduced to a series of pre-designed blockages and flows and the city is a resource through which movements and circuits of either Law or profit are to be reinforced or realized. For Kimura, Black "can't tell right from wrong": the moral codes functioning to reproduce dominance and hierarchy. For Fujimura, Black's question of "Why's the Rat back in Treasure Town?" is not even given response. Instead, he lectures Black: "I feel for you, not having parents. But you can't keep living like this…Society has rules, you know? (You have to follow them. If you want, we can get you jobs). It's important to –". Black scoffs at this lecturing and physically cuts Fujimura off mid-sentence as he kicks Fujimura's cigarette from his fingers, then responds: "We don't wag our tails for anyone!"

Immediately preceding and precipitating this encounter, Black and White come upon the detectives sitting by a fountain engaged in a conversation about the character of cities. In a moment of nostalgic reflection, Fujimura recounts to Sawada:

> **Fujimura:** Back when I was in uniform, this town had some warmth to it. Now it's stone-cold.
> **Sawada:** Cities have been cold since Hammurabi built Babylon.
> **Black:** Nebuchadnezzar II built Babylon.

This matter-of-fact declaration by Black of an historical state of affair is much more than mere pedantic correction. Through the cultural personae of Hammurabi and Nebuchadnezzar II the narrative speaks to two very different perspectives of sociality and city life taken by Kimura and the detectives on the one hand and Black and White on the other: for Hammurabi, the promulgator of Divine Decree, sociality and city life were to be ordered and hierarchically ruled through a series of eternal and transcendental codes given to him by God – Laws to be inscribed and permanently set in stone pillar; for Nebuchadnezzar II, the destroyer and builder of cities, sociality and city life were an ebb and flow, a rise and fall, a yin and yang, a current and countercurrent of power and whose dream of a giant felled tree impelled his own descent into a seven year madness and served as a lived example that "the law of arborescence" (DELEUZE and GUATTARI 1987, 292) always gives way to divergent, rhizomatic forces – even though heaven ruled it only did so through the immanent and emergent forces of everyday life.

The great philosopher of the immanent forces and physics of everyday life, SPINOZA (Ethics, III, XIII, Exp.), tells us that, just like Black and White, hope and fear cannot exist without one another – and, indeed, the entire narrative of *Tekkonkinkreet* can be understood as moving through the reciprocal principles of their urban manifestations. There is something to Black and White and their yin and yang manifestations of hope and fear that go beyond any closed binary system – the affirmative necessity of others, the collective relation and realization of desire, the empowerment of a positive sociality and need for community; but so too is there something in the fear and disempowerment of separation, the apprehension of destruction, the horrors of displacement facing Treasure Town and the (be)coming violence imposed by Snake's plans – that tell us about the city as an open yin and yang meeting place between the individual and the social: between the alleged agency and subjectiveness of autonomy and the supposed fixed frames of concrete structures (see AMES 1983, 32).

In distinction to any Levi-Straussian-like system of binary opposition and contestation between closed groupings or reifications of structural organization and hierarchical order, binaries of yin and yang are less fixed warring antagonisms than open complimentary oppositions, more transitive forces and movements between than sedimented structures or dominations across: they are synergetic relations in which there are no absolutes and they intone together with DELEUZE and GUATTARI (1987, 41) that though "[t]he word 'structure' may be used...it is an illusion to believe that structure is the earth's last word." Each principle of yin and yang is extracted from "the Tao, or Great Ultimate," (HOOKER 1999) that is the universe, and is always produced in tandem with an oppositional term – e.g. female/male, inactive/active, submission/domination, completion/creation, heaven/material, darkness/light. No principle can dominate eternally because not only is each one open to the inherent oppositional conditions of change by its very creation through the terms and presence of its compliment, but change itself is an indelible part of any phenomenon's given condition; "since...one principle produces the other, all phenomena have within them the seeds of their opposite state" (HOOKER 1999).

In this way, it would seem that the dialectical metaphysics of yin and yang bypass Cartesian isolations and closures of mind and body by working through a system where "the world of man and the world of nature constitute a seamless whole, governed by reciprocal relationships" (KITAGAWA 1968, 49). So, too, would it seem that the relationship of yin and yang is not one based on negation or separation but an always (un)folding mutual presupposition: "There is separation and unrelatedness in the Western perspective [of opposites]. Whereas, the Chinese view opposites as evolving and cycling. There is neither right or wrong, but rather there is balance, transformation, interaction, and dependent opposition" (168 Feng Shui Advisors). Yin and yang may thus be best understood not as oppositions of *opposites* in irreducible states, but the affirmative conditions of mutual transformations of life's relational becomings along and through the creational and controlling phases of the *Wu Xing*, or the five phases/elements/movements of fire, earth, metal, water and wood – the elemental ordinates of immanent de- and reterritorializations of everyday life where "[p]erson, community and nature are regarded as coextensive and correlative" (AMES 1983, 32). However, there remains a problem in all of this. DELEUZE and GUATTARI (1994, 91) explain that:

> In a sort of to-ing and fro-ing, Chinese thought inscribes the diagrammatic movements of Nature-thought on the plane yin and yang; and hexagrams are sections of the plane, intensive ordinates of these infinite movements, with their components in continuous and discontinuous features. But correspondences like these do not rule out there being a boundary, however difficult it is to make out. This is because figures are projections on the plane [of immanence], which implies something vertical and transcendent.

Perhaps in being true to its principles of yin and yang, the early history of Taoism is expressive of such a tension between immanence and transcendence, horizontality and verticality, integration and solitude: is the way of the universe to be discovered through an escape from society, as Yang Chu advocated; or is it to discover the laws and physics of all life and "live in conformity with them," as Mao Tzu taught? Or, as Chuang Tzu believed, is it by escaping to "a higher point of view," even while advancing humans' "unity with the universe" as "the unity of all things" (KITAGAWA 1968, 56–57)? While Taoism is often contrasted with the humanism of Confucianism, its social discipline and managed economies (NOSS 1974, 286), and represented as advocating an escape from 'civilization' through isolation and individualism, CLARK (1983, 71) points to a more nuanced sociality of a central text of the Tao (which intriguingly parallels MARX in many ways): "Civilization, in identifying the self with social status (citizenship, class membership, property ownership, functional role, etc.) reduces the organic social self to a narrower individual ego. The Lao Tzu [a central Taoist text named after its author] looks backward to the primordial unfragmented society and its social self, just as it points forward to a restored organic society and social person." Because Taoism approaches the world as a "matrix of relationships" beyond any individual ego, "[a]ll particulars are mutually defining and mutually determining, [and] the liberal concept of person as a locus of inalienable rights is, on this understanding, inappropriate" (AMES 1983, 32). Thus, movements away from and oppositions to social order and civilization (read: the city) are not because society is something inherently to flee, but for the production and practice of a better communal tomorrow; for the creative constitution of shared *rights* by "a people yet to come" (DELEUZE and GUATTARI 1987, 345; GARDNER, current volume).

Power as Right

While there has long been talk of rights in and to the city, *how* such rights are composed or constituted is little explored (see PURCELL 2002; PLYUSHTEVA 2009). Perhaps it is because "the concepts circulating [around rights] are individualistic and property-based...where the rights of private property and the profit rate trump all other notions of rights one can think of" (HARVEY 2008). Or perhaps, more fundamentally, it is because rights are often conflated with or left to matters of Law or morality, to seemingly detached codes swirling in realms of transcendent regulation deemed preexistent, inviolable and eternal – *a priori* matters which need not be negotiated or produced but only discovered and implemented (see MITCHELL 2003; LEFEBVRE 2008). This is symptomatic of an approach to life that understands the sociality of law and rights negatively and abstractly: "negative...[because] it sees law as a contrivance designed to limit harm and preserve a closed set of rights....abstract because it refers to rights independently of concrete situations" (LEFEBVRE 2008, 55). Through such abstraction and negation, the creativeness of the social, the positivity of communal becomings, the differentiating and inventive power of a people yet to come is buried in reified rules and bounded by categorical lines. The power of life becomes closed off, segregated, something to be shielded and protected – it is not to be lived ethically or collectively through its emergent and relational encounters but restrained and segmented into 'proper' places by way of pre-established moralities, closed doctrines, limiting discourses, reified ideologies and projected codes. The material and imaginative potentialities of life are mistaken for what is already possible and the concreteness of life's emergences are filtered into already-constituted groupings and boundaries; the vitality of life thus becoming a negative condition to be preempted and restrained.

In all of this, power is understood as fixed and absolute, autonomous and institutionalized; what Antonio NEGRI (1991, 2004), following SPINOZA, calls *potestas* (Power), or a "centralized, mediating, transcendental force of command" (HARDT 1991, xii).[3] This is the Power claimed by the tyrant, by the sovereign, by the patriarchic, by *Leviathan*, where the freedom of the multitude is to be limited and forsaken for the sake of economic order and absolute control. This is the Power sought by the Boss and Snake, whose power-play in and over the city is to be realized through unencumbered profits of a symbolic economy let loose on the one hand, and in the name of the Divine through an exclusionary and violent appropriation of space on the other. But NEGRI, continuing with SPINOZA, points to another form of power; a constitutive, affirmative and creative power: *potentia* (power), or "the local, immediate, actual force of constitution" (HARDT 1991, xii). This is a transformative power, a collective power; the power (*potentia*) of nature, of community, of the *multitudo*: of Black and White.

Right as Power

"The Rat's back," Black tells Gramps as they, together with White, bathe in a public sauna. Gramps stares at Black as he methodically washes his hair. Gramps knows what the Rat is a symptom of, and he knows what is coming (there is a force much darker and more menacing to Black than we yet know). Gramps is a homeless veteran who is blind in one eye and "[c]an't even make a living exchanging bottles." While walking the streets his requests of "Help a veteran?", "Spare change?" and "Can you help a guy get a drink?" are variously met with shrugging indifference, cruel derision or blatant disgust; he can only respond to jeers of "Stinky old man," "Don't touch me freak" and "Gross!" by slowly taking in a deep, melancholic breath and relieving it with a resolved and deliberate sigh. Even though Gramps is socially neglected, marginalized and shunned, he does not condone violence and refuses any money Black offers him because he knows it is gained through theft and violence – "I don't want your blood money."

---

3   It should be noted that even MITCHELL (2003, 22), in his description of "rights" as "at once a means of organizing power, a means of contesting power, and a means of adjudicating power, and these three roles frequently conflict," is drawing a separation between power (*potentia*) and rights. Rights are never a *means of* or *for* any function of power, but necessarily *presuppose* power (*potentia*) for their functioning.

 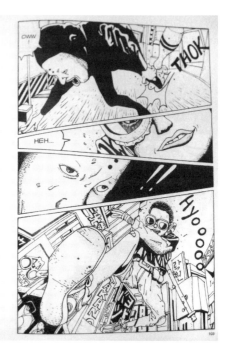

In a scene in the manga, Gramps is accosted by a local gang of fascist theocrats in numbered black bodysuits who are as much an embodiment of the current urban social order of things and its negative and closed notion of rights/power as a symptom of SPINOZA, REICH and DELEUZE and GUATTARI's (1983, 29) ongoing political philosophical question of "Why do men fight *for* their servitude as stubbornly as though it were their salvation?":

> **1:** Bumming Money off kids…What a Bastard!
> **2:** I want to say "pathetic."
> **Gramps:** You're right…Forgive me…
> **1:** What are the three duties of a good citizen?!
> **3:** Taxes! Labor! Education!
> **1:** Freeloaders like you undermine our nation.
> **3:** I want to say, "Get a job!"
> **Gramps:** Now you are going too far! How I choose to lead my life is none of your business!
> **3:** I want to say, "Don't think you're off the hook yet." You've earned Divine Retribution.

Black and White are protective of Gramps – of what may be called his "right to the city" (LEFEBVRE 1996) – and catch sight of the goings-on in the street down below. They come crashing down on 1, 2 and 3, thrashing them to bloody pulps. Following their own form and provision of retribution, Black and White create the right to the city for Gramps through their own collective force; at this moment illustrating that in claims of 'right' against 'right' it is always "force [which] decides" (MARX 1906, 259); force and desire intersecting to define the capacity and trajectory of a given (individual or collective) power. Black turns to Gramps and asks:

> **Black:** When did you get back Gramps?
> **Gramps:** Oh, about three days ago, I won't thank you. I don't approve of violence…The scent of blood in this town stings my nose.
> **White:** Are we getting a scolding?
> **Black:** "To survive in this city, you have to be strong." That's what Gramps told us.
> **White:** He did, didn't he?
> **Gramps:** …

Gramps' mute response is a sign that his idea of what constitutes strength is different than Black's. For Black, strength is physical (and violent) force and the existing social order is merely an exclusionary and hierarchical formation to sub- vert, exploit and make use of at will – the city is *his* to control: "Don't worry. I'll protect White. And our town too." For Gramps, strength is a quiet resilience in the face of the impending doom and corruption of the city: "Impossible. This town's days are numbered. At my age you can sense these things. White knows it too. That's his gift." Though he is marginalized and excluded by the existing order and balance of things, Gramps does not actively challenge the social, legal and eco- nomic circumstances that condition – and in many ways determine – his situation.

It is the violence forthcoming through negotiations of physical strength and violence over ownership of and exclusion from the city that Gramps foresees in Black's declaration that the Rat is back; and at the sauna Black's anger solidifies around an associated object of focus: "His guy Kimura is having his way with my town," Black says. Black's attention to Kimura is well placed: the Rat has ordered him to "spread some (rat) poison around here (for the brats)"; he is to be the stick of affection to the city's bee hive of affects. Gramps knows this and responds with admonition and concern to Black's declaration of ownership of Treasure Town: "Watch what you say with the 'my town' talk. It's a bad habit…This isn't any- body's town." In intensive ebbs and flows of affect, Black's relationship to the city is becoming displaced and Kimura is the most direct agent of displacement; Black's balance of life has shifted towards sadness and anger through the presence and affections of other bodies, especially Kimura's. Black knows that the right to the city, and the right to stay put in the city, for him and those he cares about (White and Gramps) can only be accomplished through a violent fight: all of this pointing to how the ebbs and flows of urban life "are not free to roam where they will" (AMIN and THRIFT 2002, 26).

Whose City is it?

Snake, Kimura, the Apaches, Black – all claim ownership over Treasure Town, and through negotiations of physical power and violence they concretely work to enforce it. Kimura tricks Vanilla into providing him with background information on Choco by leading him to believe that he would be initiated into the yakuza in return. He then proceeds to cut off Vanilla's ears – undoubtedly a response to his earlier jeer that Kimura "Got no ears! Or balls!" – and delivers them to the rest of the Apaches in a small box. He orders Choco to meet him at his office later that night. The scene cuts to Black and White sitting high atop the city and we hear Gramps' earlier dim prognostication on the outlook of Treasure Town: "This town's days are numbered. White knows it too." As Choco slowly opens the box revealing Vanilla's cut off ears, White begins to bawl inconsolably: "What's wrong?" asks Black, as he attempts to comfort White by leaning in and putting an arm around his shoulder; as the embodiment and yang of hope, White's power is that he can sense the city's intensive movements of affect.

Upon Choco's arrival at the office, Kimura reveals that he has brought Choco's mother to town, who he not-so-subtly threatens to throw off the roof of a hotel. "This is my town!" Kimura growls at Choco, knowing full well that his tactical negotiations of fear and physical violence have displaced Choco's sense of ownership over the city and have given him the upper hand of control. Meanwhile, Black has caught wind of what happened to Vanilla and sneaks out without White to confront Kimura and his gang of yakuza. He scales the façade of the building and peers in from the ledge. He stands tall with arms folded and a pitch-black raven, an oracle and symbol of death, on his shoulder. As the yakuza Yasuda walks over to confront him, Black smashes his metal pipe through the window, violently knocking Yasuda to the ground and radiating an infectious mix of fear and anger throughout the room:

> **Choco:** Black, get out of here, man. This isn't your problem! I'm fine. Look.
> **Black:** What makes you think I'm here to help you?
> **Kimura:** You've got some nerve coming here, Black.
> **Choco:** Leave him alone.
> **Black:** This is my town.
> **Unnamed Yakuza:** "Your town?" You're some comedian!

Black's anger intensifies and he shouts "Shut up!" to the yakuza's sneer while simultaneously lunging and pummeling him over the head: language being the

material and non-representational force intensifying bodily affects (see MONTAG 1999; CURTI 2009). Black and Kimura then trade violent, bloody blows: Kimura socks Black in the face and kicks his head against a desk; Black kicks Kimura in the crotch and leaps up, hanging on him and gnashing his ear; Kimura flings Black into a glass case full of alcohol bottles and then approaches him with a dagger unsheathed. "Spare the rod, spoil the child," Kimura mutters as he menacingly walks towards Black. "Stupid yakuza," Black responds with a wry grin after spitting out a shard of glass piercing his bloodied tongue. Kimura reacts, screaming "You're mine now…You're dead," as he rushes at Black. Black picks up a whisky bottle and holds it out, "This is the good stuff?" he sardonically asks before hurling it at Kimura's face, bloodying his nose and dazing him. Black then uses his metal pole to strike Kimura to the ground and the camera cuts as he is forcefully bringing down his pole to violently greet Kimura's face; the room is left with Choco weeping and all three yakuza cut, bloodied and knocked-out cold on the floor amidst the strewn wreckage of the room (the next time we see Choco he is with his mother on a train heading out of Treasure Town: "I guess this town was just too much for me"). The camera then cuts to a wreck-

ing ball demolishing a building followed by shots of the wider destruction of an entire neighborhood, coupling the preceding violent intensities of affection and affect to the material negotiations of ownership and desire extensively (un)folding in and through the city. The camera then returns us to Kimura's office where we find the three bandaged yakuza, the Rat and the angry and incredulous Boss: "Unbelievable. All you had

to do was scare some kids. Some timing! Right before introducing our new partner." Written on the wall in the yakuza's own blood is Black's unambiguous message: "MY TOWN" – the Boss' weapon of affective displacement in the fight for ownership of the city has (temporarily) backfired. In the currents and countercurrents of desire, power and control, it is Black who currently dominates the right to the city.

The Fight to Stay Put

A small abandoned car under a bridge by an industrial river serves as Black and White's home. Their recourse for ownership of this space and the exteriorities of the city that find connection in and through it is not based on transcendental codes of ownership or Law, but on ambits of control secured through their collective power impelled by movements of want and desire. Indeed, as the agents of the 'Law in the books' (see CRAINE and CURTI 2009) Fujimura and Sawada do not have the ability to protect Black and White nor do they posses the capacities of power to create the right to the city for the multitude they are meant to protect in the face of Snake's violent and exclusionary appropriation and consumption of space. Black and White's right to stay put can only be accomplished through their own fight to stay put negotiated through a microphysics of power and micro-politics of desire (the two in effect being the same thing; DELEUZE 1995, 86). It is how these micrological forces intersect in emergent and transformational encounters of affection/affect that then become primary questions of ownership and displacement, what they are and what they do.

In his introductory manifesto on "The Right to Stay Put," Chester HARTMAN (1984, 307) argues that "We need to distinguish between ownership of property that is someone's home and ownership of other kinds of property" when con-fronting questions surrounding urban transformation and the consumption of space. He maintains that this distinction must be based on people's intensive and affective connections to place (regardless of 'official' property ownership status) and the detrimental effects the severing of these connections have: "Property someone lives in but does not own engenders a special relationship between user and property – destroying bonds built up through usage (often long-term usage) – produces large individual and social costs" (HARTMAN 1984, 308) and "[i]nvoluntary residential changes also produce a considerable amount of psychological stress, which in its more extreme form has been found analogous to the clinical description of grief" (HARTMAN 1984, 306). These points of affective dis/connection are central and productively open up Hartman's notion of dis-placement as the forced movement of bodies in and from Cartesian space (see DAVIDSON 2009) to matters of desire and how it is micro-politically negotiated through different capacities of power in and through affective claims and connections of ownership. This, in turn, problematizes ideas of what constitutes ownership (see STAEHELI and MITCHELL 2008) and what counts as the home (see BOSCO and JOASSART-MARCELLI, current volume) beyond ideas of legality or structural containers and forms. MASSUMI (2002, 85) tells us that the home is "less a container than a membrane: [it is] a filter of exteriorities continually entering and traversing it." As such, the home is an intensive ordinate of dwelling and re-turn through which city life finds its most personal movements and moments (see CURTI et al. 2007); displacement then becoming as much a matter of locational movements *from* place as affective movements *in* place.

Thus, despite Gramps' admonition that "the 'my town' talk" is "a bad habit," the problem facing Treasure Town is not one of repetitious sentiment or "talk" of ownership, nor is it its implicated "habit" of approaching space as exclusionary openings and selective closures. This is not, in itself, the impetus for opposition and violence. Rather, it is the dismissal of intensive and connective affectivities and homogenization of all talk of ownership under a general moral or universal rule of Law or judgment, evading the "very different power-geometries, [the] very different geographies of power" that imbue "the social relations through which…spaces, and…openness and closure, are constructed" (MASSEY 2005, 166). Thus, the issue is not that of "bad habit[s]" but of habitual codification and reification: of form, of function, of thought, of power, of identity, of morality, of relations – as DELEUZE (1994, 5) explains, "it is useless to point to the existence of immoral or bad habits: it is the form of habit…which is essentially moral or has the form of the good" and that seeks a repetition without difference and which tells us little of concrete capacities, immanent conditions and emergent ethics of affection/affect and their corresponding increases and decreases in power that limits understandings (sometimes purposively, for exclusionary interests) of the truly relational and political natures of (urban) life.

A CITY YET TO (BE)COME

As the narrative of *Tekkonkinkreet* unfolds, the persona of the Rat, and his willing function as an ominous sign and symptom of violent urban change, begins to slowly break down. He cares about Treasure Town. There is an intensive connection. There is something emotional and personal to it that the impending changes disturb; "Treasure Town's different from the others," he tells us, "A never-never land if you will." He is upset by the "Foreigners" coming in and taking over; "One thing I know," the Rat says in response to the Boss' declaration that "This project [Kiddie Kastle] is co-financed by a 'foreigner'," "is that de-velopment doesn't suit Treasure Town." In reaction to the Boss and Snake's speech on the changes planned for the third district, the Rat interjects:

> **The Rat:** Boss.
> **The Boss:** What now, Rat?
> **The Rat:** Genpachi's strip joint is in the third. For 50 years, boys in this town have been made men there. Why tear it down and replace it with a kid's playground?
> **The Boss:** Ever a nostalgiste, eh Rat? They got more strippers than customers these days.
> **The Rat:** We can't forget the people who love and were raised in Treasure Town. (It's the people who live in this city who should shape it.)
> **(Snake:** That kind of old-fashioned provincial thinking hinders city planning and develop-ment.)
> **The Boss:** (That's right, Rat. Think globally, not locally.) Get with the program! Herr Snake is our friend and our guest. We've got lots to learn from him. And we will. So, gentle-man…[the Boss says while raising a toast]…To new beginnings!

For the Boss there is nothing and no one to forget because the multitude is not remembered. As BERGSON (1991, 137) states, "My present is that which interests me, which lives for me, and in a word, that which summons me to action" – and it is "unencumbered profit" that presently has a hold of the Boss' attention and is his summons to action; for the love of money and want of Power the multitude who love Treasure Town must be dis(re)membered. For the Rat, what cannot be for-gotten are those who have affective connections to and emotional geographies of Treasure Town. It is the inevitable violent disconnections and displacements from it that is the source of his growing despair. But, where does the Rat's concern and sentiment come from? Why is Treasure Town so unique compared to the un-named and unknown other urban spaces and places that the Rat has undoubtedly helped violently transform in the name of exclusionary control and profit before? After all, by the mere force of his presence he can "change the entire personality of a city." Why is *this* town, *this* memory of urbanity, different? Because, though he is one of the few central characters to never claim it directly, Treasure Town is *his town*.

In response to a statement by Kimura that, "I believe in nothing," the Rat remarks: "Nothing, huh? At least believe in love Kimura. Love is all you need." Treasure Town is the thing the Rat loves; it is the thing the Rat needs; and the violent changes it is undergoing – to which he has contributed no small part – is leading him into despair: "Each to his own and every man for himself. This town could care less about us. And it'll keep changing. I learned all I know on these streets. Booze, cigarettes, gambling, women, the easy life. I loved this town." It is this lack of hope and feeling of isolation in the face of urban transformation and change that leads to the Rat's death at the hands of his lifelong pupil.

Following his thrashing at the hands of Black, the Rat tells Kimura to take a break: "Consider it a paid vacation." Kimura is crushed. He is being displaced from his position in the yakuza. Following his protests and after the news finally settles in, his composure is betrayed by a trembling face: all he can do is meekly say "Mr. Suzuki…." as he stands amidst the ongoing demolition and wreckage of an entire neighborhood: different worlds of intensity intersect in their effects – displacement works in a multitude of ways. Snake seizes this opportunity to manipulate Kimura to come to work directly for him. He senses a danger in the Rat and his transformation through a strange admixture of love and despair.

Snake tricks Kimura into thinking the Rat is working with the police and then later orders him to kill his mentor: "He taught me everything I know," Kimura angrily protests. "So touching, especially with you becoming a daddy," Snake retorts, "Just imagine your lady with her stomach ripped open." Kimura's desire is pulled in different directions, but it is his role as a yakuza that ultimately wins out: life must be lived according to a code that transcends life itself.

The Rat is wise. He knows the city, he knows the streets, he knows this game, and he knows what is to come. In the end the Rat's death is a manipulation of his own working: it is less a murder than a suicide by another's hand – he has become affectively/effectively "defeated by causes external" (SPINOZA, Ethics, IV, Prop.

20, Schol.). Facing the trembling gun of a sentimentally rambling and weeping Kimura, the Rat offers him some last advice: "My final tip, be quiet when you whack someone…I taught you better than that, yes?…Don't leave prints when you ditch the piece…And burn your clothes…Most important, love your wife and child…Because love is all you need." Soon after Kimura pulls the trigger, Black finds the Rat alone and bleeding-out on the ground: "Black, this is where we all end up. You can't choose the life. It chooses you." Black picks the Rat's fallen cigarette up and places it back in the Rat's mouth as his life and its consequences flash before our eyes. Invisibly hanging in the night sky is MARX's (MARX and ENGELS 1978, 595) sentiment that: "Men make their own history, but…not…under circumstances chosen by themselves, but under circumstances directly found, given and transmitted from the past." Moments before shooting, Kimura nostalgically relates to the Rat: "You've been like a father to me"; the Rat responds: "The sins of the fathers….", but he leaves the words unfinished. The sentiment is clear.

Unlike the Rat, Kimura does not love Treasure Town. As a father (figure), the Rat has simply been an opener and closer of doors, a connector and disconnector of zones (DELEUZE 1997, 61–62). In the end, Kimura has his own passions and desires that have pulled him here and there. His earlier stoic declaration that he believes "in nothing" has dissipated under the pressure of a growing love: for his baby, for his girlfriend, and, only now does he truly begin to realize, for the Rat. He returns to Snake only to finally be qualitatively transformed by the cold detachment of Snake's reaction to the death of his 'muscle,' three alien killers brought in to rid the city of Black and White once and for all:

Snake: Damn. Faulty parts. Totally replaceable.
Kimura: Replaceable?
Snake: I call the office and order spares. A part wears out and you replace it. That's business.
Kimura: Faulty parts.
Snake: Kimura, gimme a light!

"Light this," Kimura says, pointing his gun at Snake. "Kimura, who put you up to this?" Snake coldly asks; "Love. Love and truth," Kimura replies before squeezing the trigger. The incremental realities of the cold logic of pure capitalism have finally shocked Kimura into qualitative change. He has become worthy of action.

Kimura seeks refuge from his action's consequences by taking his pregnant girlfriend and fleeing to someplace other. As he quickly packs the car he eagerly says to her:

Kimura: Let's live in a town by the ocean. Where we can hear the waves. A hammock on the porch. I can play catch with the kids on the weekends.
Girlfriend: You're acting kind of weird today. Scary.

Before Kimura can respond a motorcycle ridden by two yakuza tears around the corner, their identities clearly marked by their dark sunglasses. Kimura is shot dead and falls against the backdrop of an imagined future where he, his girlfriend and his child are playing in the shallow water where the beach greets the ocean:

Kimura: Makoto. Name the baby that. Makoto…meaning "truth."
Girlfriend: Never. I'd never have a boy.

While the commentary on the cold and detached masculinist nature of the violent fight for economic, social and physical control playing out in Treasure Town is clear, Stuart J. AITKEN (2002, 121–22) explains that such "narrative resolutions" may also point to

> a solution that leaves behind the repressed eroticism of the city and finds resolution in the country, the seaside and nature. We are left with a disturbingly normalized and naturalized…heterosexual geography that leaves behind but does not wholly dislocate not only the terror and horror [of the city], but also the law of the father. But in gaining the country/seaside/nature, space is once more produced under the tyranny of three intersecting lines of power that begin with masculinity and align with the bourgeois family and capitalism.

Of course, the fact that Kimura's narrative resolution only realizes a dream of ocean refuge in death points to the failure of all of this: the repression of difference, the futile and artificial separation of nature and the city, the unbounded rule of capitalism and patriarchy, the tyrannical production of exclusionary space; all are conditioned by the endless, differentiating cycles and causes and effects of the politics and ethics of becoming – the idiotic Power (*potestas*) and stasis of the "law of the father" always giving way to the immanent power (*potentia*) and rhythms of the laws of life.

Black and White: The City

Like Kimura, White seeks refuge from the violence of the city at the ocean, and invariably each aftermath of intense violence affectively leads his imagination there. When he is confronted by the Dusk and Dawn brothers who come to Treasure Town to instigate a turf war – "We're taking over here. Gonna kick the Cats out and rule Treasure Town" – White responds with, "So you're bad guys are you?" – White knowing that good and bad are relative terms unfolding through what a particular action is doing and for whom. Following a chase over and through Treasure Town's streets and buildings, Black and White corner Dusk and Dawn at a clock tower, who begin laughing hysterically before beating the brothers senselessly. White is then transported to the sea: "I know this place!" he excitedly says to himself as he envisions Black swimming with the fish in the rhythms of ocean water in tranquil bliss.

Sometime after their pummeling at the hands of Black and White, we learn that Dusk and Dawn were displaced from Yutaka City by "[Assassins]...Three big guys. Aliens. They ran us out of town." "Are they tough," Black asks. "Killing machines," Dusk replies. Black's face slowly transforms into a menacing grin; "Only the Minotaur is stronger," the brother says;

> **Black:** That's a good one. The Minotaur.
> **Dusk:** You don't believe in the Minotaur? The crazy kid only believes in violence.
> **Black:** I heard of him. Got a head like a cow. Total bullshit.
> **Dusk:** I don't believe in him but when I look at you I think…

Meanwhile, White offers the money in his piggy bank to Dawn: "We was gonna ride in an airplane. But you can have it," but it is initially refused: "You keep it. Take care of it," Dawn says – his most immediate concern is for White's well-being:

> **Dawn:** Forget him.
> **White:** You mean Black?
> **Dawn:** He's just looking for trouble. You're comin' with us, White. You aren't like him. Right?
> **White:** I'm alright. Bye-bye.
> **Dawn:** White.

White soon eyes the three alien assassins who displaced Dusk and Dawn from Yutaka City. They are now in Treasure Town, the "muscles" to Snake's "brains" who together "cooperate in pursuit of one goal...We work together to create a perfect piece of art." As Snake sends them off with a command in an alien tongue, Kimura anxiously asks: "What's happening? Where are they headed?"; "Cat extermination," Snake gleefully proclaims. At this moment we find White and Black together spying on the goings-on in Snake's panoptical lair from afar:

White: Three strange-looking guys. Big ones. Real scary. Curly eyebrows.
Black: Curly eyebrows?
White: They were huge!
[Black recalls the Brother's warnings: "The three assassins will come here too."; "Killing machines."]
Black: Right. They do have curly eyebrows.

White soon becomes affectively welled up. He clutches his stomach: "Black. I gots…a bad feeling" – as the affective barometer of the city, White is intensively displaced by a bodily anticipation of the anger, hatred and violence to come.

"I hate it…I hate it…I hate this town.…I hate it…I hate this town, Black. I…just thought of something real smart!...Let's build a house. By the sea. Just the two of us. And, and, we can paint it blue," White lethargically mutters to Black on top of a train as they are being hunted down by one of the aliens. After escaping near death at their car, they are on the run. Hope is leaving White: fear is taking over. Although it appears that the sea White imagines (like Kimura) is an escape to somewhere else, a someplace other, it is in actuality a "movement in place" (DELEUZE and GUATTARI 1987; DELEUZE 2003a), or the workings of the embodied imagination becoming-other – a micro-physical and micro-political constitution of power and desire creatively escaping the city to concretely reterritorialize it into a city yet to come. DELEUZE (2003b, 14) points out that

> Ocean and water embody a principle of segregation such that, on sacred islands, exclusively female communities can come to be…After all, the beginning started from God and from a couple, but not the new beginning, the beginning again, which starts from an egg: mythological maternity is often a parthenogenesis. The idea of a second origin gives the deserted island its whole meaning, the survival of a sacred place in a world that is slow to re-begin.

It would be a mistake to understand the "female" here in DELEUZE's thinking as a closed gender term. Rather, it presents the movement of *becoming* itself: "It is perhaps the special situation of women in relation to the man-standard that accounts for the fact that becomings, being minoritarian, always pass through a becoming-woman" (DELEUZE and GUATTARI 1987, 291). Becoming-woman, becoming-minoritarian, becoming-other: all are suggestive of movements away from hierarchies of Power (*potestas*), reified codes of Law, the transcendence of thought and reason and the fabricated idiocy of the 'law of the father' and its manifestation from the sickly symptoms of the closed fabulations of psycho-analysis, and movements towards the Tao, or "a field of immanence in which desire lacks nothing and therefore cannot be linked to any external or transcendent criterion" (DELEUZE and GUATTARI 1987, 157); Taoist sociality, ocean and water, the deserted island: the Outside planes of immanent becomings.

White embodies this plane; he seeks nothing external to life nor appeals to transcendence. As the alien assassin is about to hack Black with his sword, White pours gasoline over him from high above and says to the universe as if in a trance: "Agent White, Planet Earth here. I have come to defeat the alien…Agent White had a dream. Of the sea…the sky…the trees…and the wind." He then lights a match and drops it on the alien, setting him ablaze. "Let it burn…all of it," White says laughing, "Let it burn…all of it…Let it all burn. This whole city's going to burn"; "White," says Black quietly, the darkness within tempered by his love for White: fire as both destructive and creative immanent agent.

In the opening of the film we overhear Gramps speak of fire:

> What is it about the fire? So calm and peaceful…but inside, all power and destruction. It's hiding something. Just like people do. Sometimes you have to get close to find out what's inside. Sometimes you have to get burned to see the truth.

The truth of the fire is the truth of affect: the truth of how hope can become dominated by fear in the face of displacement, separation, segregation, domination and control; and the truth of how desire is transformed through the passions by violent negotiations of intensive and extensive affections. So, too, is it the truth of the city's propinquity, where "terms like 'far', 'deep' and 'distant' are replaced by rhythms which fold time and space in all kinds of untoward localizations and intricate mixtures" (AMIN and THRIFT 2002, 47). These truths come to consume Black as he and White become separated from one another.

A second alien has hunted White down and stabbed him through the stomach. Before the alien can finish him off, he is confronted by Fujimura and Sawada and flees. Bleeding profusely from his belly and half-dazed, White makes his way back to the car under the bridge – that intensive ordinate of return that serves as his and Black's home. Unable to find White in his search throughout the city, Black has been anxiously waiting for him there. He exits the car to find White near death. Even though Black told White to "stay put" in the hidden safety of a park after escaping the first alien, the car was White's original destination of return: he had a dream that sparked an emotional memory that impelled the movement of a chance trajectory that crossed his path with that of the alien's; it was a dream of an apple tree: "My apple tree!" Gotta water it. Agent White is a very smart boy!"

Earlier at the sauna, White was eating an apple and was dawned with an epiphany: "A seed….I just gots a great idea! A seed grows up to be a tree, right? And fruits grow on trees, right? We can go plant it right now!" Like the sea, the tree for White is not an escape from the oppressiveness of the city to a liberating nature, but a transhuman transformation and reconstitution of both. His desire is not for the *tree* itself, its sedimented roots and arboreal shoots, but for its coming fruit – a growth and nourishment of life that depends on rhizomatic deterritorializations of sensuous intensities (sweetness, brightness, color, texture, scent, etc.) for adventitious (re)birth and growth: White is the planter of gardens, the builder of cities, the agent of change – White is Nebuchadnezzar II.

But White is also Adam and Eve, the victim of the apple's sensuous affections. "We can't go home," Black tells White, not because of any dictate transcending their circumstances (the fruit of the tree itself is not forbidden) but because he knows the aliens – the agents of the Serpent ("Snake")[4] – will be there looking for them. Like Adam and Eve, there is no defiance of God (or Nature)[5] – there is only opposition to the attempted composition of relations-to-be imposed by Snake "who acts as a self-declared agent of God to force people out" (MCCULOCH 2007): displacement as a physical truth in the name of economic growth as Divine Right; the poisoned apples sold through the sermons of false prophets/profits. There is no tree of knowledge of Good and Evil: Black knows there are no such transcendental codes; there are only good and bad affections and affects negotiated along and through affirmative lines of desire and power as right – Black is SPINOZA (so, too, is he NIETZSCHE).

---

4   The character's name in Japanese, *Hebi*, is translated as "Serpent" in the manga.
5   As DELEUZE (1981) explains working through SPINOZA: "When God reveals to Adam that the apple will act as a poison, he reveals to him a composition of relations, he reveals to him a physical truth and he doesn't send him a sign at all."

The City: White and Black

The Power (*potestas*) of Divine Right and false prophets always gives way to the power (*potentia*) of creativity and the *multitudo*: the power of life. Black is this power, he embodies it, embraces it and consumes it; until it consumes him. At the hospital Black will not leave White's side until he emerges from the darkness: "This place...I know this place" White says as he dreams of the ocean, and awakens. Black separates with White, leaving him first at the hospital and then with Fujimura and Sawada. His active purpose in life is to protect White, and outside forces are passively eroding his ability to do so. With his head in his arms Black tells Gramps: "I'm so tired, Gramps. That took a lot out of me. White's...different. All he does is talk silly talk. I don't know if I can protect him anymore...I don't believe in God or anything...But it seems like God is trying to take White away from me." Gramps: "That boy's stronger than you think. And you're not as tough as you think. My eyes tell me he's been the one protecting you, am I wrong? I believe in you, Black. Even if God looks the other way." Black can simply respond with a hint of a grin as a tear rolls down his cheek.

The scene shifts to Fujimura and Sawada removing White from the hospital. They are confronted outside by Black who, to White's incredulous rage, allows them to take him away. Their separation delves Black first into fear and sadness, and then into madness. Snake: "[T]hat brat continues to be a nuisance...People say he's changed." Kimura: "Unbalanced since losing White. Out for blood." It is Winter, the domination of yin. Black is on a rampage, the alien assassins are on the prowl, and Snake is on the brink of frustration: "This is my town. I need to keep my streets clean." Black crosses paths with the Rat, who ominously relates to Black his own sense of displacement and hopelessness just as Black does to the Rat his own coming death:

> **The Rat:** The word on the street is that you're possessed. The more you love something, the deeper the pain. Treasure Town will never be what it was, Black. Even if we stop, the city keeps going.
> **Black:** You're going to die. It's written on your face.
> **The Rat:** [Laughing] Then I'll see you in hell. I'll be waiting there for you.
> **Black:** This town is hell.

The Rat understands the dual nature of the affect of love – "For even as love crowns you so shall he crucify you. Even as he is for your growth so is he for your pruning. Even as he ascends to your height and caresses your tenderest branches that quiver in the sun, so shall he descend to your roots and shake them in their clinging to the earth" (GIBRAN 1997, 5) – the Rat is Khalil GIBRAN (so, too, is he SPINOZA). Black understands that "hell is itself the effect of an elementary injustice, an avatar of elements" (DELEUZE 1997, 46) and finds that "[e]very one who has ever built anywhere a '*new heaven*' first found the power thereto in his *own hell*" (NIETZSCHE 2007, 119) – Black is NIETZSCHE (so, too, is he D. H. LAWRENCE).

In the compositions of effects and affects of the injustices and elements of a violent and greedy urban transformation, Black is becoming memoryless with no hope or love of the past or future as the hell of displacement, separation and the absence of White affectively transforms him:

> **Black:** Have you seen white? I told him to stay put.
> **Gramps:** What're you going on about? White is with the police. I told you, Black.
> **Black:** I told him to stay put. Damn.
> **Unnamed Homeless man:** Poor kid. Needs a shrink.
> **Gramps:** I doubt a doctor could understand his pain. Nobody can understand his pain.

Gramps understands the singular nature of affect and the emergent meanings to which it gives rise: "What matters…is not the meaning of life in general, but rather the specific meaning of a person's life at a given moment" (FRANKL 1992, 113) – Gramps is Viktor FRANKL. White senses the intensive growth of sadness and fear in Black as he furiously scribbles his marker on pieces of paper; "Is this a cow," asks Sawada. "Minotaur…He's going to eat up this town. I gots a bad feeling," White replies. Earlier, White has told Sawada of his and Black's indelible connection and need for one another:

> **Sawada:** Hey, White. Why don't you ever ask about Black. Aren't you worried about him?
> **White:** White is missing lots of screws.
> **Sawada:** Missing screws?
> **White:** Yeah. I need screws for my heart. God made me broken.
> **Sawada:** Broken?
> **White:** Broken. Black too. He's broken. He's missing screws too…for his heart.
> **Sawada:** Black's broken too?
> **White:** Yep. But I gots all the screws Black needs. I got every one. I got every one!

The distilled darkness dwelling within Black violently emerges as he is confronted by the two surviving alien assassins inside of Kiddie Kastle. Black is carrying a doll that his madness has made him believe is White. One alien pulls out a shotgun and blows the doll's head half off. Black screams in terror and his anger and rage are unleashed. Meanwhile, White is shouting and crying in hysterical fits. Fujimura and Sawada have no clue what is happening. With crossbow arrows sticking out of him, Black is transported to the tranquility of the sea – "This place…Okay, yeah, alright. I know…Know this place. I know this place." He is at peace with the specter of death hanging over him. He is at peace because he is in White's place. At this moment one of the aliens puts his blade to Black's forehead, drawing a stream of blood, and the Minotaur swoops down from high above:

> **Minotaur:** Don't fear the darkness, Black. Behold…the true power…of the darkness.

The aliens are soon dead at the hands of the Minotaur and Kiddie Kastle is on fire and crashing down as a result of one of the alien's errant missiles. Black trembles and recoils in fear at the presence of the Minotaur:

Minotaur: Don't be scared. Follow me. I'll show you the way. I'm here with you, Black. [The Minotaur hugs Black deeply.]
Minotaur: This is the landscape of your creation. A deeper, higher dimension: Darkness.
Black: Darkness?
Minotaur: Don't fear it. It transcends all.
Black: Transcends all?
Minotaur: The power of darkness.
Black: [In front of the ocean:] White, is that you?
Minotaur: No, don't get clouded. Do…not…get…clouded. That leads to death.
[White screams out and flails uncontrollably.]
Black: Is this a dream?
Minotaur: More real than reality…Free the darkness in you.
Black: White?
Minotaur: Show him your true power.
Black: My…true power?
Minotaur: Let go. Darkness is truth…C'mon, Black.
Black: Who are you?
Minotaur: Lost between Light and Dark you found another part of yourself.
Black: [Recognizing a gaping cut on the Minotaur's hand as his own:] That wound?
Minotaur: [Now appearing without a mask:] Let's do it! Now we can really bathe our town in blood!!
[The camera cuts to the scene of Kimura's death.]
Black: This is my power?
Minotaur: This is just the beginning. You're capable of so much more.

The camera quickly cuts to a montage of gambling, isolation, drugs, rape, murder, suicide, greed, and all around bloody hatred, violence and death, before bringing us to Fujimura trying to wake White from his daze:

Fujimura: White. Hey.
Sawada: Wait. You want to see him, right, White? You want to see Black? [Sawada hugs White deeply, his frigidness transformed to warmth through White's affections of hope and love.]
White: [Jumping back from the memory of an airplane flying over his and Black's home:] I gots all the screws Black needs! I've got every one!

Minotaur: White again?!
White: Stop!
Black: White?
Minotaur: Always getting in the way! He's a hypocrite. A pretender. I'm not just a shadow cast by your light. I'm the real you. Darkness is true, transparent, beautiful. Embrace solitude. [Black is watching a white dove furiously flap its wings in the darkness and hears White: "What do you see over there? What do you see over there?"]

Black's memories return: his love for White, their friendship and need for one another, their strength together, the first time they met; the truth of fire/the truth of affect:

White: Be happy be happy.
Black: Be happy be happy.
Minotaur: Not him. You need me! What do your eyes tell you? What do you truly believe in?
Black: I…I….I believe in…White.
Minotaur: [As Black descends from the darkness and reterritorializes with the open immanence of the world:] Look at the scar on your hand. It's there to remind you. I'm there, inside you. Always. Whenever you need me.

Only together can White and Black find an active balance and composite power (*potentia*) to overcome the Minotaur; that devourer of humanity and dweller of the labyrinth, whose complex closed-ness masks its linear and single path: the dark logic of capital laid bare. But theirs is not a binary or homogenizing togetherness, fixed balance or unchanging power. They depend, too, on the love and actions, passions and desires, forces and powers of Gramps and Dusk and Dawn, Fujimura and Sawada, and even Kimura and the Rat, to overcome the sadness and violence imposed by Snake and the Boss – a (be)coming community always building across and between differences through the mutual interests of "a generative multiplicity of divergent and discontinuous lines of flight with their own spaces and times" (AMIN and THRIFT 2002, 28). The death of the Rat at the hands of his own self-imposed isolation and hopelessness and the death of Kimura in his attempts to flee to someplace other speak to the fact that only together, through an affirmative togetherness, through a collective power and creation of this place as someplace-other, can there be a (re)founding of rights; ones built on emergent socialities of care, needs and encounters of mutual reciprocation and interests instead of transcendental codes, Laws and rules of exploitation (normalized and veiled under a 'natural' life of "competition"). This comes not from outside society, but (un)folds the Outside of society – the everyday place of creation and the new (DELEUZE 1988b) which shows us that there is another way: only collectively can the multitude be the slayer of the Minotaur and the founder of the city – only together are they Theseus.

CONSIDERATIONS: A CITY AND A PEOPLE TO BECOME

Tekkonkinkreet illustrates that to speak of urban transformation and change with no attention paid to affections and affects is not only to miss out on the most forceful relations of the most personal geographies of the body and the home but, perhaps, the most pronounced immediacy of city life. Further, it provides power-ful insights into how affective and personal negotiations of passion and desire in the city intersect and (un)fold in particular contexts of ownership to concretely speak to wider events of urban transformation and change by providing "localized expressions of endo-events and exo-events, the 'inside-of' and 'outside-of' force relations that continuously enfold the social sites they compose" (MARSTON et al. 2005, 426). Movements of violence and displacement in and from Treasure Town come both from the inside and the outside: the Boss, the Rat and Kimura are its agents from within; Snake, the alien assassins and Dusk and Dawn are its agents from without – all of this pointing to how "couplets" of local/global meet up in creative ways through the everyday processes of city life to become "mutually constituted" (MASSEY 2005, 185).

While there initially appears to be a level of xenophobia or provincial localism working through *Tekkonkinkreet* by way of its fearful portrayal of "foreigners," "aliens," and urban change, the very syncretic nature of Treasure Town and the fact that Black and White open up to (and open up) the Dusk and Dawn brothers and Sawada undermines such an argument: in this city belonging or not belonging and matters of urban change as good or bad are questions of whether one is attempting to assert domination and control along lines of exclu-sionary Power and desire (i.e. this is what constitutes "the bad guys" for Black and White), never matters of identity or authenticity defined through points of representation. Indeed, it is only Black and White as a multiplicity in communion together as marginalized embodiments of hope and action, love and joy, power and practice that can heterogeneously work to constitute a different people, a different community, a different city to (be)come. In this way, *Tekkonkinkreet* goes beyond the horizons of many sci-fi/fantasy films by directly illustrating the "detrimental consequences" of urban transformation and change in "the contexts of the various marginalized and excluded who actually live in…cities…[and t]hat the forces making the urban landscape are multiple and not necessarily focused in the hands of one almighty father figure" (AITKEN 2002, 118). Intersecting the heterogeneous, emergent and relational movements of power and desire in *Tekkonkinkreet* with HARVEY's (2008) argument that "[t]he right to the city is…far more than a right of individual access to the resources that the city embodies: it is a right to change ourselves by changing the city more after our heart's desire," points to how difference itself, in all of its multiplicitous and heterogeneous phenomenality, connects not through pre-formed or pre-designed groupings of identity, but on personal experiences of activity and passivity formed by and through the *pre-personal* forces of affection and affect.

By (temporarily) slaying the Minotaur, by collectively becoming Theseus, do the people to come in Treasure Town create a trajectory for a Treasure Town to come through the social as more-than-human (WHATMORE 2002). CROOK (1998, 524) explains that

> The contemporary myths of the Minotaur are those of the everyday as the source and site of unity, of resistance to strategic power and of the inexhaustible vitality of life. These myths, along with the theses of 'taken-for-grantedness' and 'living history', converge on a defence of the *purity* of the social. To press the conceit to its limits, the Minotaur-like monstrosity of the everyday serves as an alibi for the purely social character of social relations, a defence against the charge of truly heterogeneous monstrosity. The myths of the everyday ensure that organic processes, technical artefacts, bodies, texts, weather patterns, musical sounds and all other threatening legions of otherness remain safely outside the 'social'.

It is in this way that *Tekkonkinkreet* displays its truly heterogeneous monstrosity: by becoming a fantastically materialized Spinozan-Deleuzian progenitor and buggery-mate of "immaculate [re]conception" (DELEUZE 1995, 6) pushing bodies to act and thought to think differently about what *counts* as the everyday. It was SPINOZA whose creative and bold challenges to orthodoxy and dogma were declared "deeds" so "monstrous" by the *ma'amad* (Sephardi community leadership) in 17TH century Amsterdam that he found himself religiously and socially excommunicated and spatially exiled (see KASHER and BIDERMAN 2001). And it was DELEUZE (1995, 6) who ass-fucked "from behind" both his enemies and loved-ones to bare "monstrous" children, an act for "cloning a corruptible genetic code in order to frustrate the endless reproduction of the same" (DOEL 1999, 35) through a serenely violent (pro)creation very likely gleaned from SPINOZA's own seminal penetration and transformation of Descartes' body of thought – each an unpure sexual act in their own way by way of a productivist "excessiveness" that pushes "the development and transformation of a species" (GROSZ 2008, 33) along different social, cultural, and bodily lines of flight to becoming-other (see DELEUZE and GUATTARI 1987).

Like SPINOZA and DELEUZE, *Tekkonkinkreet* tells us that there is no purity of the social, only multiplicity and (ad)mixture – only with the heterogeneous com-

positions of the ocean and the trees, the wind and the sky, the earth and the fire –
 only with and through the microphysics and ethology of the living city and a
"landscape of life" (CURTI 2008b) – can we become the monstrous sowers of a
loving warmth and toilers of a shared social care of difference countering the
monstrous coldness and social displacements of capital with the fruits of love and
joy – only together are we Johnny Appleseed. So, too, in the multiplicitous forces
of our displaced togetherness do Black and White tell us that here are no guaran-
tees, there are no absolutes – "Look at the scar on your hand. It's there to remind
you. I'm there, inside you. Always. Whenever you need me" – there are only
forces and capacities of desire and power enhanced or diminished along and
through different compositions and relations of affection and affect.

All of this tells that it is only through *a hope which forcefully acts* that
difference in and of the city can become along alternative lines and rhythms: "for
there to be *change*, a social group, a class or a caste must intervene by imprinting
a rhythm on an era, be it through force or in an insinuating manner" (LEFEBVRE
2004, 14). Only with White is Black's darkness and fear deterritorialized by the
lines and rhythms of the imagined ocean of love and hope, and only with Black is
White able to be (re)territorialized along lines and rhythms of understanding and
the empowering organization of knowledge. As "power implies knowledge as the
bifurcation or differentiation without which power would not become an act"
(DELEUZE 1988b, 39), it is only through multiplicitous movements of hope and
need, knowledge and care continually and collectively acting *together* in a community of mutual interests to materially overcome passions of fear and sadness
that the power and the capacity of the city can be enhanced. "[T]here is neither
hope without fear, nor fear without hope" (Ethics, III, XIII, Exp.) – nor is there
light without darkness, nor darkness without light. White is the multiplicitous
hope for the happiness of a better tomorrow, Black is the knowledge that this hope
will never be realized without an active fight for the perpetual today; together they
are the power of multitude whose hopeful anger is a life-affirming gift – "everything seems pregnant with its contrary" (MARX and ENGELS 1978, 577) is no
longer a negative dialectic: MARX becomes Taoist.

REFERENCES

AITKEN, S. C. (2002): Tuning the self: city space and SF horror movies. KITCHIN, R. and J. KNEALE (Eds.): *Lost in Space: Geographies of Science Fiction*. London, 104–122.

AMES, R. T. (1983): Is Political Taoism Anarchism? *Journal of Chinese Philosophy* 10, 27–47.

AMIN, A. and N. THRIFT (2002): *Cities: Reimagining the Urban*. Malden.

ANDERSON, B. and A. HOLDEN (2008): Affective Urbanism and the Event of Hope. *Space and Culture* 11 (2), 142–159.

BERGSON, H. (1991): *Matter and Memory*. New York.

CLARK, J. P. (1983): On Taoism and Politics. *Journal of Chinese Philosophy* 10, 65–88.

CRAINE, J. and G. H. CURTI (2009): A(u)tuando o Rio: a lei, o desejo e a produção da cidade em *Tropa de elite*, de José Padilha. *Pro-Posições* 20 (3), 87–108.

CROOK, S. (1998): Minotaurs and other Monsters: 'Everyday Life' in Recent Social Theory. *Sociology* 32 (3), 523–540.

CURTI, G. H. (2008a): From a Wall of Bodies to Body of Walls: Politics of Affect | Politics of Memory | Politics of War. *Emotion, Space, and Society* 1 (2), 106–118.

CURTI, G. H. (2008b): The Ghost in the City and a Landscape of Life: A Reading of Difference in Shirow and Oshii's *Ghost in the Shell*. *Society and Space* 26 (1), 87–106.

CURTI, G. H. (2009): Beating Words to Life: Subtitles, Assemblage(s)capes, Expression. *Geo-Journal* 74 (3), 201–208.

CURTI, G. H., J. DAVENPORT and E. JACKIEWICZ (2007): Concrete Babylon: Life Between the Stars. To Dwell and Consume with(in) the Fold(s) of Hollywood, CA. *Yearbook of the Association of Pacific Coast Geographers* 69, 45–72.

DAVIDSON, M. (2009): Displacement, Space and Dwelling: Placing Gentrification Debate. *Ethics, Place and Environment* 12 (2), 219–234.

DELEUZE, G. (1981): DELEUZE / SPINOZA, Cours Vincennes (Lecture) – 13/01/1981 http://www.webdeleuze.com/php/texte.php?cle=34&groupe=Spinoza&langue=2 (accessed September 26, 2010).

DELEUZE, G. (1988a): *Spinoza: Practical Philosophy*. San Francisco.

DELEUZE, G. (1988b): *Foucault*. Minneapolis.

DELEUZE, G. (1989): *Masochism: Coldness and Cruelty*. New York.

DELEUZE, G. (1990): *Expressionism in Philosophy: Spinoza*. New York.

DELEUZE, G. (1994): *Difference and Repetition*. New York.

DELEUZE, G. (1995): *Negotiations 1972–1990*. New York.

DELEUZE, G. (1997): *Essays Critical and Clinical*. Minneapolis.

DELEUZE, G. (2003a): *Francis Bacon: The Logic of Sensation*. Minneapolis.

DELEUZE, G. (2003b): *Desert Islands and Other Texts 1953–1974*. Los Angeles.

DELEUZE, G. and F. GUATTARI (1983): *Anti-Oedipus: Capitalism and Schizophrenia*. Minneapolis.

DELEUZE, G. and F. GUATTARI (1987): *A Thousand Plateaus: Capitalism and Schizophrenia*. Minneapolis.

DELEUZE, G. and F. GUATTARI (1994): *What is Philosophy?* New York.

DOEL, M. (1999): *Poststructuralist Geographies: The Diabolical Art of Spatial Science*. Edinburgh.

ESTREVA, G. (1992): Development. SACHS, W. (Ed.): *The Development Dictionary: A Guide to Knowledge as Power*. London, 6–25.

FRANKL, V. E. (1992): *Man's Search for Meaning*. Boston.

GIBRAN, K. (1997): *The Prophet*. Hertfordshire.

GROSZ, E. (2008): *Chaos, Territory, Art: Deleuze and the framing of the earth*. New York.

HARAWAY, D. (1991): Situated Knowledges: The Science Question in Feminism and the Privilege of Partial Perspective. http://science.consumercide.com/haraway_sit-knowl.html (accessed May 7, 2010).

HARDT, M. (1991): Translator's Forward: The Anatomy of Power. Antonio NEGRI: *The Savage Anomaly: The Power of Spinoza's Metaphysics and Politics*. Minneapolis, xi–xvi.

HARTMAN, C. (1984): The Right to Stay Put. GEISLER, C. C. and F. J. POPPER (Eds.): *Land Reform, American Style*. Totowa, 302–318.

HARVEY, D. (2008): The Right to the City. *New Left Review* 53. http://www.newleftreview.org/?view=2740 (accessed September 25, 2010).

HOOKER, R. (1996): Chinese Philosophy: Yin and Yang. http://www.wsu.edu/~dee/CHPHIL/YINYANG.HTM (accessed September 23, 2010).

KASHER, A. and S. BIDERMAN (2001): Why Was Baruch De Spinoza Excommunicated? GENEVIEVE, L. (Ed.): *Spinoza: Context, sources, and the early writings*. London, 59–99.

KITAGAWA, J. M. (1968): *Religions of the East*. Philadelphia.

LEFEBVRE, A. (2008): *The Image of Law: Deleuze, Bergson, Spinoza*. Stanford.

LEFEBVRE, H. (1996): *Writings on Cities*. KOFMAN, E. and E. LEBAS (Eds.). Oxford.

LEFEBVRE, H. (2004): *Rhythmanalysis: space, time, and everyday life*. New York.

MARSTON, S. A., J. P. JONES III and K. WOODWARD (2005): Human Geography without Scale. *Transactions of the Institute of British Geographers* 30, 416–432.

MARX, K. (1906): *Capital: A Critique of Political Economy*. New York.

MARX, K. AND F. ENGELS (1978): *The Marx-Engels Reader*. Tucker, R. C. (Ed.). New York.

MASSEY, D. (2005): *For Space*. London.

MASSUMI, B. (2002): *Parables for the virtual: Movement, Affect, Sensation*. Durham.

MATSUMOTO, T. (2007): *Tekkonkinkreet: Black & White*. San Francisco.

MCCARTHY, G. (1985): Marx's social ethics and critique of traditional morality. *Studies in Soviet Thought* 29, 177–199.

MCCULOCH, J. (2007): My Life is Choked with Comics #11 – Tekkonkinkreet (aka: Tekkon Kinkreet, aka: Black & White). http://www.savagecritic.com/jog/my-life-is-choked-with-comics-11-tekkonkinkreet-aka-tekkon-kinkreet-aka-black-white/ (accessed September 25, 2010).

MITCHELL, D. (2003): *The right to the city: social justice and the fight for public space*. New York.

MONTAG, W. (1999): Bodies, Masses, Power: Spinoza and his Contemporaries. London.

NEGRI, A. (1991): *The Savage Anomaly: The Power of Spinoza's Metaphysics and Politics*. Minneapolis.

NEGRI, A. (2004): *Subversive Spinoza: (Un)Contemporary Variations*. Manchester.

NELSON, R. H. (2001): *Economics as Religion: From Samuelson to Chicago and Beyond*. University Park.

NIETZSCHE, F. (2007): *The Genealogy of Morals*. Stilwell.

NOSS, J. B. (1974): *Man's Religions*. New York.

168 Feng Shui Advisors: Yin and Yang Theory. http://www.168fengshui.com/articles/yin-and-yang-theory/ (accessed September 23, 2010).

PLYUSHTEVA, A. (2009): The Right to the City and the Struggles over Public Citizenship: exploring the links. *Urban Reinventors Online Journal* 3 (9) 1–17.

PURCELL, M. (2002): Excavating Lefebvre: The right to the city and its urban politics of the inhabitant. *GeoJournal* 58, 99–108.

SASSEN, S. (1996): Whose City Is It? Globalization and the Formation of New Claims. *Public Culture* 8, 205–223.

SHERINGHAM, M. (1996): City Space, Mental Space, Poetic Space: Paris in Breton, Benjamin and Réda. SHERINGHAM, M. (Ed.): *Parisian Fields*. London, 85–114.

SPINOZA, B. (1996): *Ethics*. New York.

TEKKONKINKREET. DVD. Directed by Michael Arias. [United States]: Sony Pictures Home Entertainment, 2007 (original film release 2006).

STAEHELI, L. A. and D. MITCHELL (2008): The People's Property? Power, Politics, and the Public. Abingdon.

THRIFT, N. (2007): *Non-representational Theory: Space, Politics, Affect*. London.
WHATMORE, S. (2002): *Hybrid Geographies*. London.
ZUKIN, S. (1995): *The Culture of Cities*. Malden.

Stuart C. Aitken

THE MOST VIOLENT VIBRATIONS BETWEEN HOPE AND FEAR [1]

> That a class which lives under the conditions already sketched and is so ill-provided with the necessary means of subsistence, cannot be healthy and can reach no advanced age, is self-evident...The manner in which the great multitude of the poor is treated by society today is revolting. They are drawn into the large cities where they breathe a poorer atmosphere than in the country; they are relegated to districts which, by reason of the method of construction, are worse ventilated than any others...All conceivable evils are heaped upon the heads of the poor...They are exposed to the most exciting changes of mental condition, the most violent vibrations between hope and fear; they are hunted like game, and not permitted to attain peace of mind and quiet enjoyment of life (ENGELS 1845, 107–9).

The plight of urban poor – from their housing conditions and quality of life to their displacement at the whim of planners and policy makers – is conditioned by struggles for basic needs. The 'right to stay put' as a basic need is questionable, but there is no doubt from ENGELS' famous diatribe that from the beginnings of industrial capitalism, the poor are shunted into landscapes that do not serve their basic needs. This essay is about progressive social justice and why rights-based arguments for decent housing and living conditions are problematic at best. Rather, by focusing on ENGELS' 'violent vibrations between hope and fear' I suggest that, first, abstract rights can never trump actual, complex emotional experiences and, second, even when a right is implicated it may produce no determinate consequences because of local geographic contexts. Both these engagements with rights derive from leftist Harvard Law Professor, Mark TUSHNET's (1984, 1363) critique of the liberal theory of rights as "a major part of the cultural capital that capitalism's culture has given us." The first issue relates to geography's enduring focus on affects and emotions and, from this, the second issue raises the specter that most bureaucratic decisions are either irrational or, at least, indeterminate. The issue becomes one of immediacy: the fight for, rather than the right to, decent housing.

Before the term 'gentrification' entered the academic lexicon, there was recognition by observers from ENGELS onwards that working class struggle is born out of violent exchanges between hope and fear, where people

> are deprived all enjoyments except that of sexual indulgence and drunkenness, are worked every day to the point of complete exhaustion of their mental and physical energies, and are

---

[1]   A small portion of the material in this essay showed up in a Scottish Geographical Journal article on children and Scottish film entitled "Poetic Child Realism: Scottish Film and the Construction of Childhood" (2007, 123, 1, 68–86).

thus constantly spurred on to the maddest excesses in the only two enjoyments at their command.

And if they surmount this, ENGELS (1845, 110) goes on to say, "they fall victims
to want of work in a crisis when all the little is taken from them that had hitherto
been vouchsafed them." As an example of the loss of the vouchsafed (read rights)
I use the Glasgow neighborhood of Maryhill as portrayed in *Ratcatcher* (1998), a
fictional movie by celebrated Scottish photographer and filmmaker Lynne
RAMSAY. RAMSAY presents a story of quiet desperation at the height of the 1974
dustbin workers' strike, a time when ENGELS' "ill-provided landscapes" find sharp
relief. To accentuate the somber desperation, RAMSAY's movie is scored lyrically
with the vibrant (but also violent) voices and perspectives of young people and the
numbed bewilderment of their caregivers. I focus on violent vibrations between
hope and fear through James, the young protagonist of RAMSAY's narrative, and
his father George.

If there is hope in this movie it comes from resilience and other internal energies. I note, from the work of Scottish psychiatrist R. D. LAING (1960), that there
is an important connection between institutionalized social repression and individual psychic repression. LAING argued that ontological insecurity finds form
when society shackles human desire and takes away individual freedom in a
'double-bind' where contradictory societal messages lead to a series of self-de-
structive sequences. This connection between society and individual repression
may be understood in a more hopeful way through DELEUZE's notion that desire is
always positive. In opposition to a Freudian or Lacanian lack, DELEUZE saw desire as a social formation – an infrastructure – that builds the potential of
becoming other. Lack, for DELEUZE, appears only in the sense that the infrastructure within which we invest our desire in turn produces that lack (DELEUZE and
GUATTARI 1977; SMITH 2007, 74). This coincides with, and extends, LAING's
concern with the disparaging relations between social and personal repression and
puts it on a road where desire and happiness coincide. From LAING and DELEUZE,
then, I argue that emotions and feelings power transformative possibilities. I note
how the infrastructure in RAMSAY's *Ratcatcher* continuously negates the
expressed feelings of James and George. I note further how the infrastructure
finds its fascist form in the everyday social formations of families and the ways
they are embedded in everyday street-life (see FOUCAULT, In DELEUZE and
GUATTARI 1977, xvi). I look at the tension between fear and hope, between over-
coded urban spaces and spaces of desire. To these points, RAMSAY unabashedly
tackles the contexts of children's desires – including sexual desire – in the midst
of abstract representations that seek a sanitized and compliant working-class. I
look at the ways that RAMSAY portrays Maryhill and the changing mental
condition of James and his father, whose sentiments "vibrate between hope and
fear."

HOPE AND FEAR IN OVER-CODED URBAN SPACES

Phil HUBBARD (2006, 97) points out that the practical turn in urban theorizing has lately involved a re-engagement with the work of Marxist humanists of which ENGELS was undoubtedly the first. Henri LEVEBVRE moved ENGELS' sentiments forward with a clear articulation of the way urban space is produced (1991) and that there exist people's rights to the city (1996). Arguing that the laboring body is central to the way cities work, LEVEBVRE noted that bodies have the capacity to create cities in which a wide range of desires are realized. LEVEBVRE's humanist hope was that people create cities that fulfill their bodily needs and desires, but his Marxist fear was that abstract representations deployed by architects and planners over-coded spaces of everyday practice to create a de-corporealized city (HUBBARD 2006, 103). In what follows I look at the ways Maryhill is created as an over-coded de-corporealized space of capitalism and the ways that RAMSAY's Images, narrative and cinematic practices construct spaces of desire.

Located to the north-west of Glasgow, Maryhill is a traditionally working-class neighborhood that owes its origins to Hew Hill, the Laird of Gairbraid. In 1763, Hill left his estate to his penniless daughter Mary who founded coalmines in the hopes of generating funds. The mines were too wet to be profitable, but fortunes changed in the 1780s when the Forth and Clyde Canal was cut through the area so that it could connect with the River Kelvin. The construction involved several locks and the Kelvin aqueduct. A village grew with the building of a dry-dock boatyard and the commerce drawn to the area by the canal. Land from the Gairbraid estate was bequeathed for more development on the condition that the village was called Maryhill after Hew's forlorn daughter. By 1830, the area attracted more boat-building, a sawmill and an iron-foundry, which helped support a population of 3,000. It was absorbed into the city of Glasgow in 1891.

By the mid-twentieth century, Maryhill comprised a number of traditional middle-class sandstone tenements and large Victorian town houses as well as accommodations to support the working class population. In 1954, Maryhill was designated as one of 23 Comprehensive Development Areas (CDAs) proposed for peripheral Glasgow. Peripheral housing schemes within the CDAs were to provide around 27,000 houses for families displaced from areas in the inner city designated as slums. The displacement was called "directed migration" (GIBB 1983, 159) and was established to diminish the intense concentration of working class families within the city. Densities of 163 people per acre in Glasgow's inner city were two or three times higher than other industrial cities in Britain in the 1950s. Between 1951 and 1961 outmigration from the city averaged over 20,000 people per year as growth in redevelopment schemes such as those established in Maryhill gathered momentum. The majority of development took the form of three- and four-storey tenements, with scattered single-family homes, terraced homes and some high-rises. The tenement facades hid a diversity of different housing types and some fairly high residential densities.

Desperate to provide housing for families displaced by the destruction of inner-city housing, the new areas were developed with virtually no services or space for small businesses. GIBB (1983, 166) notes that for the first few years in some of these peripheral estates, school children had to walk to collection points where they were bussed back to the schools in the inner city. In some cases, new tenement houses were temporarily converted into primary schools for the younger children.

> Planners ideals of fresh air and natural green recreation space mean little to those above primary school age, and from the first arrivals in the 1950s, a kind of 'culture shock' gripped many families removed from the comforting familiar, sooty womb of tenement slums (GIBB 1983, 166).

The choice of staying put in high-density inner-city tenements did not materialize as city planners deployed their large-scale abstract representations, which involved the wholesale demolition of inner slums to coincide with the so-called comprehensive peripheral development. Many of the multi-storey towers and terraced housing built in the 1960s quickly degenerated to a worse condition than the original inner-city tenements and in places the supportive infrastructure was tenuous at best. The 1974 dustbin workers' strike accentuated an already deprived landscape.

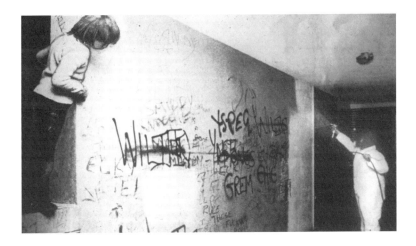

EMOTIONAL SITES

RAMSAY was born in Maryhill on December 5, 1969. As a child she witnessed the outcome of directed migration from the inner city. Her childhood experiences in Maryhill color a number of her photographic and filmic projects, with *Ratcatcher* (1998) quickly gaining critical acclaim for its compassionate and sometimes brutal portrayal of young people and their place-based sensitivities. With her non-professional actor children, RAMSAY creates despair, hope, transcendence and poetic lyricism. Places and their ontologies are an important part of this creation. In coining the term, Theodore SCHATZKI (2005, 467) argues that *site ontologies* maintain that "…human coexistence is inherently tied to a kind of context in which it transpires." The contexts involved are known as sites, and they comprise assemblages in which some of what occurs or exists in them are inherently parts. For SCHATZKI, a site is a place or location or building, but it can also be an institution or an event.

Out of site ontologies, RAMSAY's characters and narrative evolve. The opening sequence of *Ratcatcher* looks out of the window of a tenement at the Forth and Clyde canal. [2] The sense from the beginning of the movie is of a part of Maryhill that is socially tight-knit, with frustration and a violent edge bubbling just under the surface. Materially, it is clear that the community suffers from dilapidation and a lack of infrastructure. The streets are filled with garbage, but they are also animated by children at play and adults going to and fro. Along with the canal, Maryhill's high density 1960s corporation housing provides a *mise-en-scene* for the narrative.

In *Ratcatcher,* the Forth and Clyde plays a pivotal role where children play, experiment with sex and die. It is a site ontology in the sense that it is a man-made waterway, constructed for the profit of a few industrialists while creating jobs for a cadre of working-class laborers through the middle of the twentieth century. By the 1970s it is no longer used for commerce, and the bodies of children populate its grimy banks and torpid waters. A river does not "run through" Maryhill in the sense of Robert Redford's famous 1992 Hollywood blockbuster or Clint Eastwood's *Mystic River* (see AITKEN 2006). Stagnant and polluted, the canal does not cleanse or enable forgetting. RAMSAY's Forth and Clyde canal is a repository of exploitation, repression and malignant guilt.

---

2   The canal in the movie was built specifically for *Ratcatcher* because what is left of the Glasgow canal system is too polluted for active use. RAMSAY (2002) notes that "[b]ecause we basically couldn't – for health reasons – we couldn't use the real canals which are pretty disgusting. You know, they are pretty polluted as well. You know, Jane had the bright idea, or me and Jane had the bright idea of 'let's build our own canal' you know, em, it all got a bit bonkers considering we're not like, we didn't have a massive budget, but it felt like the only way to do it. So it began to become this big deal: we started digging three feet down and we found all this toxic waste, you know, I mean it was like 'we cannae, what are we going to do now?' So we had to get some local government to come and take the toxic waste away, that had been left there for some time. That meant we couldn't go too deep. And it became a total nightmare. I think if we'd known what we were getting ourselves into we'd never have done it."

When James is involved in the accidental drowning of Ryan Quinn, he holds on to his culpability and is continually drawn back to the canal. The drowning is the culmination of excessive play, but for James emotional retribution continually re-emerges from the canal's murky depths. It is a dark, eerie connection to which James is drawn. The waters are about loss and blame; they bind a community around Mary Hew Hill's hapless attempts at generating income from her father's estate off of the backs of working-class laborers. The canal exploits and exacts a toll on its workers and their children.

Later in *Ratcatcher,* the canal is joined by a second site ontology in the form of a new peripheral housing estate discovered by James during a random bus ride. As part of his fantasy of getting out of Maryhill, the peripheral estate becomes a site of hope. RAMSAY creates a sense that James' escape is not at all about Ryan Quinn's death, but about moving away from the over-coded Maryhill to a green-field site. By the time the movie gets to James' bus ride, the garbage is piling up on the streets of Maryhill and the peripheral estate is set up as a contrast, where new appliances and wood are piled up next to the houses.

The fantasy of escape is highlighted metaphorical when Kenny, one of James' neighbors, gets a pet mouse for his birthday. When Kenny shows the mouse to James, they are surrounded by a gang of local boys, one of whom grabs the mouse, which they start throwing between them. They then throw the mouse to James and taunt him to smash it against the wall: "Let it fly James go on…Go on James, fuckin' kill it…Just fling it at the wall." James is a paragon of gentleness in the midst of this inchoate violence. He replies "the moon" to Kenny's question about where the mouse will fly. The brutality of the gang's peer pressure is contrasted dramatically with the kindness of James who gives the mouse back to Kenny: "It's done enough flying today, better put it back in its cage." Kenny responds by going up to his flat, attaching the mouse to a balloon and launching it out the window, in an attempt to send the mouse to the moon. The ensuing sequence of images is James' dream fantasy of the mouse flying up to a cheese-filled moon to join other mice in blissful community. The effectiveness of the scene works at several levels. First, it comes at 50 minutes into the movie and provides some relief from the film's tension and brutality. Second, the issue of Kenny killing his pet mouse resonates with an earlier scene when George kills a feral mouse with which James and his sister were playing. Third, it connects to James' fantasy of journeying to the peripheral housing estate because the mice's moon reflects his dream of a community of friends far away from Maryhill. As an event, the fantasy is linked to the site ontology of the green-field housing estate.

RAMSAY uses James' journey to the peripheral estate to portray a sense of freedom: her young charge looks at times bemused and at times joyous. At the estate, Maryhill's high density housing, garbage strewn streets, constant noise and every present threat of swift violence is contrasted with a relatively low-density peripheral green-field site with inside toilets and quiet.

Rather than looking out at fighting children and bickering neighbors, James looks out of a new house to a field full of wind-swept grain. The irony is that the peripheral estate is yet another planning gambit, yet another over-coded space for more directed migration. The site ontology of the peripheral estate is its allure to people trapped in the blight of Maryhill in the same way that, a few decades earlier, Maryhill and directed migration were a panacea for inner-city blight.

Violent exchanges between hope and fear guide RAMSAY's craft, and site ontologies power underlying tensions. In terms of craft, on any particular day of filming *Ratcatcher*, RAMSAY says that she was willing to let the landscapes lead her children, with some places structured as despair while others reflect hope. The peripheral estate is a site of hope, of renewal, of family:

> Well that's actually one of my favourite scenes in the film: the scene when [James] goes to the new housing estate. I think I have this thing about when I was a kid: something really beautiful about going into empty houses, or houses that have been abandoned. There is a kind of weird, kind of special feeling about who has lived there or who is going to live there. Or the newness of the place. That was kind of my favourite scene when I wrote it (RAMSAY 2002).

At the housing estate, a field is framed by a window that James then proceeds to climb out of for a beautifully emotive romp in the corn. This particular scene was improvised and, as such, it changes the energy of the film: "The shot where he sees a cornfield through the window," explains RAMSAY, "I wanted it to look like a painting. I want it to be a frame within a frame. It was already in the script. I think [James] was amazed. Why am I standing in this set in the middle of a corn-field? And that comes through." RAMSAY's eye for portraiture and still photography elicits bemusement and surprise from James. RAMSAY combines framing from photographic techniques with a strict sense of narrative devices such as cross-cuts to create emotive representations.

Ontological contrasts evolve around sexual experimentation and corporeal connections through RAMSAY's cross-cutting techniques. The site of the canal is about sexual experimentation for Margaret, a neighborhood girl who allows the gang of local bullies access to her body in a less than casual way. James befriends her after retrieving her eyeglasses, which are thrown into the canal by one of the bullies. Their friendship enables James to move through (but not beyond) the drowning and helps Margaret overcome her lack of self-esteem. The relationship that develops between Margaret and James offers hope and there is a bathroom scene where they share a corporeal but not sexual intimacy. The corporeality begins with James combing lice out of Margaret's hair and the scene moves into the bathroom for hair washing. RAMSAY connects this scene to the site ontology of the canal with a cross-cut to James' father George saving Kenny from drowning in the canal. Anguish and fear in the canal water is contrasted with affection and fun in the bath water. The tender moments that James and Margaret experience helps heal both their souls but they are ultimately insufficient to offset the larger social infrastructure of Maryhill and the violence of its environment. The last time we see her, Margaret is being led into an outside toilet by a group of boys.

DEPRIVED SITES OF HUMAN CO-EXISTENCE

The relationship between James and George is uneasy at best. James' family is in line for a move to a peripheral estate but that move is adjudicated by Glasgow City Council. An ironic twist in the movie finds stark relief when a delegation from the Council arrives to evaluate James' family's potential for directed migration. This critical scene in *Ratcatcher* brings together the fantasy world of James with the drunken realities of George and the larger social inequities and prejudices perpetrated by the representatives of the City Council. That the right to housing is adjudicated through seemingly rational decision-making is offset by bureaucrats who cannot see beyond the context that is framed before them. Once more, RAMSAY uses her experience as a photographer. "I am particularly inspired by photography, stills photography," she says, "because basically you are capturing a detail in a person that says something about their whole world at times. So, within a still there can be a lot of narratives."

The critical scene comes after George has saved Kenny from drowning in the canal and lies recovering, asleep on his bed hide-away in the parlor. It opens with James entering the flat, framed (like a still portrait) in the opening of the hide-away and (like still photography) we are drawn to what is not seen (George slumped on the bed).

In a moment that mirrors an earlier scene where he pulls his mother's holed nylon stocking affectionately over her foot, James picks up some breakfast cereal from his father's forgotten bowl and sprinkles it onto the prone form on the bed. RAMSAY (2002) talks about this small act with the mother as a rendering of James' love to the extent that the scene with the father suggests a disdainful, or at least ambivalent, relationship. A knock is heard at the door, again out of the scene, and James turns to respond.

A conversation is heard and we are treated to another classic still featuring George's leg and the kitchen framed in the door of the bed-hideway. What follows is an emotionally uncomfortable scene (I squirm at George's attempt to rouse himself and make the best of the situation) when the two representatives of the council come in with James: "We've come to inspect the condition of your property and to assess your standard of living. Have we caught you at a bad time?"

I'm Miss McDonald. This is Mr. Mohan.
We're from the council.

George is not drunk at this moment; his disheveled appearance is a direct reflection of the cereal that James sprinkled on him and the dirt from saving Kenny.
George babbles about his heroism at the canal and his frustration at the state of the
garbage-strewn streets outside the tenement. These moments of his life are lost to
the council assessors who are intent on their quick dismissal of the family's ap-
propriateness for a new house. They seem not to hear George's comments as they
make some brief notes on a clipboard. When George suggests they come back
when his wife is home, they say they have enough and take their leave.

The Gillespie family is caught like the rats in the garbage outside, and their
plight is responded to with the same insensitivity that the council later shows
when dealing with the mess caused by the dustbin workers' strike. George
struggles to inform the council representatives of his situation, but he is in one of
LAING's double-binds. The council appraisers are clearly more concerned about
making their assessment and getting out quickly. Once the inspectors leave, James
eagerly approaches his father with a question that reveals (but only to us) the de-
tails of his bus trip to the peripheral estate:

> **James**: Are we getting a big hoose wi a bathroom, a toilet, an' a field. Are we Da?
> **George**: Come here you…Can you no fuckin' do aenythin' right? Whit did you let 'em in here fir, eh?
> **James**: 'Cause a thought that…
> **George**: 'Cause you thought…
> **James**: Are we still getting' a new hoose?
> **George**: Well if we don't, it's your fuckin' fault. Noo get oot o' ma sight.

George's response reveals in part his embarrassment at the context of the ap-
praisal but also an unthinking self-centeredness wrought out of fear and frustration. It is an indication of LAING's socially stimulated internal repression.
RAMSAY's portrayal of George is not unsympathetic, but this scene and others
suggest his lack of connection to James and the rest of his family.

A little later, George is summoned before the Mayor of Glasgow to receive an
award for his heroism at the canal.

At this auspicious occasion, we are reminded of the peripheral housing estate when James' older sister asks their mother why they have not moved yet, when the family of a school-friend has just received notification that they are to move. The irony of George's bravery in saving Kenny from the canal is that how he represents himself – dirty and disheveled – to the Council representatives an hour or so later is precisely why the Gillespie's request for a displacement is turned down. The context of housing rights is displaced as indeterminate in the sense that TUSHNET (1984, 1364) means when he argues that "the invocation of rights allows us to engage in rational arguments" but deliberations over rights "falsely converts into an empty abstraction (reifies) real experiences that we ought to value for their own sake." The lack of value assigned to George's heroism is further suggested by a chaotic and banal awards ceremony where the Mayor does not get George's name right. George's response, echoing ENGELS' sensitivities to the plight of the working man, is to skip out of the ceremony as quickly as possible so he can join his pals in the pub and get drunk.

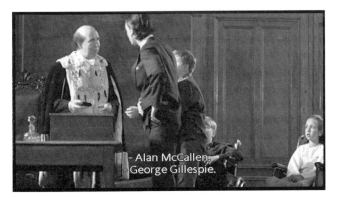

- Alan McCallen,
- George Gillespie.

Weaving and slurring on his way home after the afternoon's drinking session, George agrees to hold onto the kitten of a young girl who wants to buy an ice-cream: "Sure a can, be careful crossin' that road there." Seeing him staggering and leaning against a garbage-strewn wall with the cat, a gang of youths pounce on an opportunity: "Look at him, man, he's steamin'…Where d'ya git the cat, ye pussy?"

The consequent attack with a sharpened steel comb (an emblematic weapon of 1970s Glasgow youth culture) leaves George with a gash across his cheek. He returns home to his celebrating family bleeding, but flush from his award George has a gift for James: a pair of football boots. He finds his children laughing and cheering on James as he tries to learn how to jitter-bug with his mother. George tries to join the merriment and pushes the boots at James.

Frustrated by his inability to connect with the light-hearted reverie in the flat, George slaps his wife who gets in the way as he tries to get James to try on the boots. James retorts as he runs out throwing the boots at his dad: "A don't fuckin' like fitba."

TO FIGHT AGAINST RIGHTS

The event of the dustbin workers' strike serves as a *mise-en-abyme* for the movie in the sense that it mirrors the unattainable endpoint of desire in an over-coded urban space. Like watching a vanishing point in a hall of mirrors, the garbage piled in the streets of Maryhill offer children an opportunity to chase rats as part of dystopian desire. *Ratcatcher* is a journey through a troubled Glaswegian childhood with side-trips, often simultaneously, into brutality and beauty. The journey begins with the ontological presence of the canal and ends with the presence of the peripheral housing estate. Childhood is constructed through the sharp words and actions of adults, the viciousness of peer pressures, and the aesthetic of daydreams and fantasies. James' narrative comes to a close with a visual recognition of his aloneness in Maryhill after the strike ends. The garbage, a site ontology in itself to the extent that is a context for James' social interaction, is gone. What he is unable to communicate to his mother, father and sisters is his need for community, akin to the one Kenny's mouse found on the moon.

James is lying on the bed where he dreamt about the mice on the moon. His younger sister comes in sleepily and they cuddle. Throughout the movie they've bickered and fought for their mother's attention; this is the first intimation of affection between them. James leaves his sleeping sister and goes over to the hideaway bed where, once more, he fixes his mother's holed stocking. This place of love is contextualized by his looking out past the scene: his look is towards the canal. Cross-cut to the canal with a small, barely noticeable figure, in the distance. Cross-cut to bubbles similar to those that intimated Ryan Quinn's drowning at the beginning of the movie. As James sinks to his death we witness the final fantasy: his family and friends in a procession through the cornfield to their new house. The previously empty cornfield filled with smiling people carrying furniture contrasts the empty streets of Maryhill.

In *Ratcatcher*, places and events create site ontologies, metaphorically and politically. RAMSAY representation of 1970s Maryhill is about fantasies, insecurities, and moralities but it is also about economies of place. It is about contexts of social and spatial control and brutal peer pressures. It is at one moment stark and vicious and at another, surreal and pregnant with potential. RAMSAY raises questions about morality, participation, advocacy and their politics in place. In this regard, RAMSAY makes her landscapes work through a desire that is always positive. They are places that coalesce with characters to precipitate hope. Questions arise about cultures of poverty, about the ways over-coded places work in and through those cultures, and about the ways sex, drunkenness, violence and death are left as the only palatable ways out. Answers to these questions are not obvious or coherent: with some considerable discomfort they foment indistinctly out of the ways RAMSAY creates particular kinds of emotional filmic geographies – based upon tiny moments and small pushes – that embody social and spatial relationships in ways that other forms of film realism miss entirely. There is an aesthetic at work here that is sometimes uncomfortable, sometimes tragic, sometimes hopeful, and always poetic.

RAMSAY's future is not one of urban utopian optimism, nor is it dystopian. It does not offer material solutions. James' fantasy world of new homes, open fields, and pets that survive brutality by flying to the moon lead ultimately to his own tragic departure, courtesy of the canal. Hope, rather, lies with landscapes and places that foreground the other-side of the god-trick; not as map-like contexts of surveillance, control and escape but as characters out of and through which new politics might emerge. This moves in the direction of an entire theory of affectivity that this chapter can do little more than hint at: It suggests a purely immanent ethics that is neither an individual's conscious will nor their pre-conscious interests in the sense, say, of Marxian class interests (SMITH 2007, 74). This leads to the fundamental question of political philosophy as defined by DELEUZE and GUATTARI (1977, 29): "Why do people fight for their servitude as stubbornly as their salvation?" George's humanity saves Kenny from drowning in the canal, and his stake in the system leads to his confused embarrassment at the Mayor's annual award ceremony. The ceremony is portrayed by RAMSAY as yet another put down for George, another repression that thwarts his interests, but it is the societal basis of his desire to be a hero. The affective basis of George's humanity is the impulse that saved a drowning boy, but the infrastructure creates the moniker hero for that act, and a ceremony to fulfill a need (in all of us) to be heroes. The system introduces the lack in George that is insufficiently filled in public ceremony. James' desire for a house with "a bathroom, a toilet, an' a field" is reinforced and frustrated by a City Council that cannot clear the garbage from the streets of Maryhill or see past the dirty body of George in the bed hideaway, but also creates the seeming utopia of a peripheral, green-field, housing scheme. Desire – our drives and affects – are a complicated part of the capitalist infrastructure rather than some simple, internal reality. RAMSAY creates a Maryhill of despair but also of hope, of material degradation but also of transcendence. This is precisely the poetics and politics that her work realizes.

REFERENCES

AITKEN, S. C. (2006): Leading Men to Violence and Creating Spaces for their Emotions. *Gender, Place and Culture.* 13 (5), 491–507.

DELEUZE, G. and F. GUATTARI (1977): *Anti-Oedipus: Capitalism and Schizophrenia I.* trans. Robert Hurley, Mark Seem and Helen Lane. New York.

ENGELS, F. (1845/2009): *The Condition of the Working Class in England.* Oxford.

FOUCAULT, M. (1977): "Preface" to Gilles Deleuze and Félix Guattari. *Anti-Oedipus: Capitalism and Schizophrenia I,* trans. Robert Hurley, Mark Seem and Helen Lane, p iii–xx. New York.

GIBB, A. (1983): *Glasgow: The Making of a City.* Burrell Row, Kent.

HUBBARB, P. (2006): *City.* New York and London.

LAING, R. D. (1960): The Divided Self: An Existential Study in Sanity and Madness. Harmondsworth.

LEFEBVRE, H. (1991): *The Production of Space.* Translated by D. Nicholson-Smith Oxford.

LEFEBVRE, H. (1996): *Writing on Cities.* Translated by E. Kofman and E. Lebas. Oxford.

RAMSAY, L. (2002): Interview on the DVD release of *Ratcatcher.*

SCHATZKI, T. R. (2005): Peripheral Vision: The Sites of Organizations. *Organization Studies* 26, 465–484.

SMITH, D. W. (2007): Deleuze and the Question of Desire: Towards an Immanent Theory of Ethics. *Parrhesia* 2, 66–78.

TUSHNET, M. (1984): Essay on Rights. Part of a Symposium on A Critique of Rights. *Texs law Review* 62, 1363–1412.

Colin Gardner

CONSTRUCTING A CINEMA OF MINORITIES: 'STAYING PUT,' *MEMORIES OF UNDERDEVELOPMENT* AND THE INVENTION OF A PEOPLE YET TO COME[1]

> *Every onlooker is either a coward*
>
> *or a traitor.*
>
> *(FANON 1966, 159)*

In *Cinema 2: The Time-Image*, Gilles DELEUZE (1989, 189) constructs an ontology of the cinematic body as

> no longer the obstacle that separates thought from itself, that which it has to overcome to reach thinking. It is on the contrary that which it plunges into or must plunge into, in order to reach the unthought, that is life. Not that the body thinks, but, obstinate and stubborn, it forces us to think, and forces us to think what is concealed from thought, life.

Thus for DELEUZE, following Nietzsche, thought is thrown into the essential categories of life – i.e. the body and its postures – in each case judged by the extent of what it can do as a creative force. It's through the body, its theatrical *gest* and attitudes, that the cinema makes its alliance with the spirit and with thought. For example, "In Godard, the attitudes of body are the categories of the spirit itself, and the gest is the thread which goes from one category to another" (DELEUZE 1989, 194). This *gest* is also inextricably caught up with time, for the cinematic body includes its past and present experiences and those which are yet to come, its energy but also its tiredness, exhaustion and inertia.

But what of the body politic, the larger *figure* of the people? What of collective somatic utterance? For DELEUZE, in classical cinema, the people are most certainly present, even if they are oppressed, tricked, made subject, blind or largely unconscious. They are real before being actual, ideal without being abstract. "Hence," he notes, "the idea that the cinema, as art of the masses, could be the supreme revolutionary or democratic art, which makes the masses a true subject" (DELEUZE, 1989, 216). However, this latent cinema as a popular art was compromised first by totalitarianism and subsequently, after the defeat of fascism, by Cold War conformity: "In short, if there were a modern political cinema, it would be on this basis: the people no longer exist, or not yet...*the people are*

---

1   This essay is dedicated to the memory of Teshome GABRIEL (1939–2010), dear friend, mentor, nonwestern film scholar and activist.

missing" (DELEUZE 1989, 216). However, far from being a pessimistic prognostication of the impossibility of progressive cine-activism, this impasse creates a dystopia ripe for the re-construction of a truly revolutionary Third Cinema: "not that of addressing a people, which is presupposed already there, but on contributing to the invention of a people" (DELEUZE 1989, 217). In this regard the nonwestern film auteur – whether it be Ousmane Sembene in Senegal, Glauber Rocha in Brazil, or Humberto Solas and Tomás Gutiérrez ALEA in Cuba – is in a situation of producing utterances which are already collective assemblages because the director or writer acts as a catalyst for potential forces or incommensurable events that give rise to the *idea of a people yet to come*: "The author can be marginalized or separate from his more or less illiterate community as much as you like; this condition puts him all the more in a position to express potential forces and, in his very solitude, to be a true collective agent, a collective leaven, a catalyst" (DELEUZE 1989, 221–2).

This essay will explore this creation of a specific collective utterance through a close reading of ALEA's seminal and paradigmatic film, *Memories of Underdevelopment* (1968), itself adapted from Edmundo DESNOES 1962 novel, *Memorias del Subdesarrollo.*[2] As Michael CHANAN (1990a, 3), explains:

> Underdevelopment is an economic concept, referring to the relationship between a country with the status of an economic colony and the metropolis that colonizes it. The title therefore claims that now colonization is over. Underdevelopment has been replaced by the Revolution. But it turns out that the new title is somewhat ironic, for what the film shows is the way that people continue to carry the mentality of underdevelopment within them, how it weighs them down, and how it becomes a problem.

Sergio, ALEA and DESNOES' protagonist (played in the film by Sergio Corrieri), is a bourgeois intellectual and would-be writer in the vein of DESNOES himself who decides to buck the trend and stay behind in Havana after his wife and family have fled as exiles to Miami. Sergio epitomizes the *rentier* class: under Batista he owned a furniture store given to him by his father as well as the apartment block in which he currently lives in the Vedado district of Havana. His only source of income is the twelve years' worth of payments he receives as compensation by the Revolutionary government for the confiscation of his property. By exercising his individual right to "stay put," Sergio ends up as a case study in geographical limbo but also as the catalyst for a new form of cinematic 'ethics.' On the one hand, Sergio is too 'advanced,' too critically self-reflexive and ironically distanced to identify with the residual Third World underdevelopment that characterizes post-Revolution Cuba. On the other he is also too 'backward' (i.e. equally underdeveloped because of a lack of historical involvement) to find a satisfactory political or psychological 'place' in the emerging collective identity of Cuba as a viable 'third way' between western imperialism (represented by the U.S.

---

2 The novel was translated into English by the author and published by New American Library in 1967 as *Inconsolable Memories.* This version incorporates several sequences from ALEA's film that are not in the original Spanish release.

economic blockade) and Soviet ideological and military influence. As ALEA (1990, 210) himself puts it:

> His contrariness and the source of his dissension lie in knowing himself to be alienated by cultural patterns foreign to his own environment, and nevertheless unable to struggle to assert himself. He is already a defeated man who reveals the cultural colonization that has victimized us throughout our history, the consequence of which, within the revolution, is located in a general sense of our underdevelopment.

ALEA's film is structured dialectically via two parallel trajectories that both inflect and ultimately *infect* each other as we are forced to negotiate the creation of a new, hybrid form of de-territorialized text that defies easy resolution. For ALEA (1990, 204), "The image of reality provided by *Memories of Underdevelopment* is multi-faceted – like an object contemplated from different viewpoints." These viewpoints take the form of an interweaving between what ALEA calls "evocation" (which is largely subjective) and "representation" (predominantly objective), although the director is careful to point out that neither term should be seen as mutually exclusive and untainted by infection by the other. Thus, on the one hand we have the narrative's subjective focus on Sergio, expressed through a fractured series of semi-chronological vignettes[3] (they are undated, sketchy diary entries in DESNOES's novel) linked together through voice-over narration, distinctly scopophilic point-of-view shots, and focalized perspective through a combination of shot-reverse shot editing and *mise-en-scène* – which is represented formally and structurally by what DELEUZE (1986, 217) calls the cinematic Action-Image, the "reaction of the centre to the set [*ensemble*]." Here, centered on an acting body, the Action-Image is all motility: bodies determine place as the product of a movement through and across geographic space. Epitomized by the realist model of the Classic Hollywood Cinema, its defining characteristics are active modes of behavior taking place within determined milieux: e.g. the law of the gun vs. the Law Book in the typical western or gangster genre. Here, affects and impulses are embodied in modes of *behavior* which lead to motor action that mutually transforms both the body and the context of its action (the milieu). Thus,

> what constitutes realism is simply this: milieux and modes of behaviour, milieux which actualise and modes of behaviour which embody. The action-image is the relation between the two and all the varieties of this relation. It is this model which produced the universal triumph of the American cinema, to the point of acting as a passport for foreign directors who contributed to its formation (DELEUZE 1986, 141).

Realism is thus a dynamic, transformative form, in which both individual and context undergo change for the better and learn to co-exist for each other's mutual benefit. The most important quality of the Action-Image is that it is an entirely organic representation. It is dependent on the foundation of a large gap existing between encompasser and hero, milieu and modifying behavior, situation and action, which can only be bridged in progressive increments throughout the course

---

3   Paul SCHROEDER (2002, 26–7) notes that the film is composed of thirty-one separate sequences.

of the film. This is because the hero is rarely ripe for action; "like Hamlet, the action to be undertaken is too great for him. It is not that he is weak: he is, on the contrary, equal to the encompasser but only potentially. His grandeur and his power must be actualized" (DELEUZE 1986, 154). Realist violence is thus active, constructive, and generative of change and the building of character.

The irony in *Memories of Underdevelopment* is, of course, that Sergio, the ostensible action-hero, is largely indifferent and inert in relation to his post-Batista social surroundings, unable to act by either opposing or embracing the Revolution. On the other hand, as ALEA notes (BURTON 1977a, 9), he is also attractive, sensuous and seductive:

> In my view, the Sergio character is very complex. On one hand, he incarnates all the bourgeois ideology which has marked our people right up until the triumph of the Revolution and still has carry-overs, an ideology which even permeates the proletarian strata. In one sense Sergio represents the ideal of what every man with that particular kind of mentality would like to have been: rich, good-looking, intelligent, with access to the upper social strata and to beautiful women who are very willing to go to be with him. That is to say, he has a set of virtues and advantages which permit spectators to identify to a certain degree with him as a character. The film plays with this identification, trying to insure that the viewer at first identifies with the character, despite his conventionality and his commitment to bourgeois ideology.

As one might expect, Sergio's potential activism is displaced onto a series of ill-fated and highly manipulative affairs with women, each of whom represent conflicting elements of Sergio's personality (most specifically his abortive attempts to integrate into a series of different social realities) as well as a specific class and cultural trope within the new Cuba.

Thus, Laura, his ex-wife (Beatriz Ponchora), represents the Euro-Americanization of the Cuban petit-bourgeoisie (the class that ultimate flees the Revolution for Miami), whereby Sergio transformed her from a "slovenly Cuban girl" into an elegant woman but one who is also artificial and empty, signified by the cosmetics strewn about on her dressing table and the Park Avenue gowns left behind in her closet. Secondly there is the bohemian Hanna, Sergio's lost European love and "the best thing that ever happened in my life." A pre-war refugee from Nazi Germany, she is a natural blonde and, as the ideal woman, represents "the real thing" rather than Laura, the pale imitation. Unfortunately, Sergio let his "prize" get away: he postponed their marriage and his writing career for the sake of his furniture store. As Julianne BURTON (1977b, 18) points out, "Material aspirations, rather than more appropriately romantic obstacles, are revealed to be responsible for the loss of Sergio's one 'true love.'" Thirdly, Elena (Daisy Granados), who he picks up on the street, represents the urban working class. However, she is most certainly *not* a 'new Cuban woman' in the ideological sense and retains all the worst traits of the pre-Revolutionary petite-bourgeoisie. Indeed, she is an aspiring actress (to 'unfold her personality') with a penchant for elegant consumer goods from the U.S. While she resists Sergio's sexist attempts to shape her into a readymade European sophisticate – he takes her to art museums and Hemingway's Cuban country house (now a museum/shrine to the great man) as

ways to educate her in the finer points of cultural and literary life – Sergio casually dismisses her lively spunkiness and stubborn independence as a hopeless
inconsistency, yet another symptom of underdevelopment. This is partially borne
out when her family – led by her boorish brother (René de la Cruz) – accuse him
of raping Elena, an underage virgin. Fortunately for Sergio – and much to his
surprise – he is found innocent and released by the courts. Finally, there is Noemí
(Eslinda Núñez, star of the second episode of Solas's three-part drama, *Lucia*)
who represents the rural proletariat and in many ways manifests the reality of an
idealized Revolution. For Sergio – who sees her as exotic because of her
Protestant baptism – she embodies the innocence and hope of the new order and
the possibility of new ideological beginnings. However, unlike in DESNOES's
novel, where the affair is consummated, sex with Noemí is as far fetched as inter-
course with the image of Botticelli's *Venus* that Sergio fondles with such wishful
familiarity. After all, Noemí is the maid and is off limits purely as a matter of
class protocol. As a sex object she thus remains a pure fantasy.

The second of *Memories of Underdevelopment*'s two parallel trajectories is
the objective representation of historical culture and the collective re-emergence
of Revolutionary Cuba itself. This is rendered as a form of Brechtian
Verfremdungseffekt through the use of intertitles-cum-chapter headings (e.g.
"Havana 1961," "Noemí," "A Tropical Adventure"), inserted archive and
documentary footage of prisoners captured at the invasion of Playa Girón, photo-
essays of Sergio's childhood, everyday street scenes shot with a hidden-camera,
and the insertion of the fictional Sergio into real locations (Hemingway's house
utilized as a setting for fictive action and a stinging critique of the artist as coloni-
alist) and public events, such as a round-table on literature and underdevelopment.
The latter includes appearances by DESNOES himself (who, as in the novel, is the
target of some well-aimed self-reflexive critique) as well as the American play-
wright and author of the forward to the English edition of *Memories of Underdevelopment*, Jack Gelber. Indeed, self-reflexivity is taken to the point of comic
absurdity in a scene in the screening room at ICAIC (The Cuban Film Institute)
where an unidentified ALEA oversees the screening of semi-pornographic film
clips (former victims of Batista's censors) which repeat in a slapstick, staccato
fashion. As the lights go up, Sergio asks his friend:

> **Sergio**: Where did you get the clips?
> **Alea**: They showed up one day. These are the cuts the commission made. They said they
> were offensive to morals, good breeding, all that.
> **Sergio**: What are you going to do with them?
> Alea says he's going to put them into a film [i.e. *Memories of Underdevelopment*]: "It'll be a
> 'collage' that'll have a little bit of everything."
> **Sergio**: Will they approve it?

In addition to underlining the more progressive nature of cultural production
under the new regime (after all, the clips have been passed for exhibition within
the film we are watching), the sequence also posits an artistic alternative to

Sergio – Alea himself – as a bourgeois artist who has turned his creative energies to making a progressive work of art with great political commitment.[4] This is in opposition to Sergio's own novelistic dabblings, where, at the very end of his diaristic novella, he is forced to admit, perhaps following in Samuel Beckett's footsteps, the necessity of going beyond words into the realm of the incommensurable image, or failing that, silence:

> The October Crisis is over. The Caribbean Crisis. The Missile Crisis. To name huge things is to kill them. Words are small, meager…Want to preserve the clean and empty vision of the days of crisis. Things around me and fear and desires choke me. It's impossible. Beyond this, I have nothing to add. Finished. Man is (I am) sad, but wants to live…Go beyond words (DESNOES 1990, 175).

For ALEA, then, the film's two trajectories – evocation and representation – represent a political one-two combination punch, in which a cerebral space is opened up whereby the formation of a people yet to come may be constructed by the viewing audience. Thus the initial subjective identification with Sergio is essentially a softening up process for the more ideological, collective focalization to come:

> As the film progresses, one begins to perceive not only the vision that Sergio has of himself but also the vision that reality gives to *us*, the people who made the film. This is the reason for the documentary sequences and other kinds of confrontation situations which appear in the film. They correspond to our vision of reality and also to our critical view of the protagonist. Little by little, the character begins to destroy himself precisely because reality begins to overwhelm him, for he is unable to act. At the end of the film, the protagonist ends up like a cockroach – squashed by his fear, by his impotence, by everything (BURTON 1977a, 9).

Thus as the new revolutionary society fights for its own collective right to "stay put" and emerge from the shackles of both pre- and post-Revolutionary underdevelopment, the film's ostensible anti-hero becomes completely irrelevant (because he is himself symptomatic of a far more *retardataire* underdevelopment: a refusal to act) at film's end as the island feverishly mobilizes itself in face of the October Missile Crisis of 1962 and possible annihilation. DELEUZE's 'event that is life' is effectively manifested by a defiant Castro, as he proclaims in his television speech: "We are all one in this hour of danger, and it is the same for all of us, revolutionaries and patriots, and victory shall be for all." For all except Sergio, who remains aloof and uncommitted as he stares out across the rooftops from his verandah, destined to remain outside the groove of Revolutionary history. So where does that leave us, the spectator? For ALEA,

> the spectators feel caught in a trap since they have identified with a character who proceeds to destroy himself and is reduced to…nothing. The spectators then have to reexamine themselves and all those values, consciously or unconsciously held, which have motivated them to

---

4   In this respect, as Carolina Ferrer of the Université du Québec à Montréal pointed out to me, ALEA is the perfect representation of Che Guevara's ideal of the politically committed Revolutionary "New Man."

identify with Sergio. They realize that those values are questioned by a reality which is much stronger, much more potent and vital (BURTON, 1977a, 9).

We noted earlier that, in light of DELEUZE's taxonomy of Western cinema, the subjective, evocative side of the film's narrative register is manifested largely through the Action-Image. This trope is manifested formally through montage – specifically the (largely invisible) edit. In this case, images and intervals are linked within a narrative chain of action-reaction formation. Here the spectator is encouraged to leap imperceptibly across the filmic splice that links shots into scenes and then into larger sequences in order to construct a sense of logical movement through the perception of continuous action. In contrast, the interjection of the documentary and real-life sequences form a discontinuity and false movement more in line with DELEUZE's (1989, xi) definition of the Time-Image in *Cinema 2*. In this case

> Time ceases to be derived from the movement, it appears in itself and itself gives rise to false movements. Hence the importance of false continuity in modern cinema: the images are no longer linked by rational cuts and continuity, but are relinked by means of false continuity and irrational cuts. Even the body is no longer exactly what moves; subject of movement or the instrument of action, it becomes rather the developer of time, it shows time through its tirednesses and waitings...

Typical examples of the Time-Image are the Chronosign: "an image where time ceases to be subordinate to movement and appears for itself"; the Crystal-Image or Hyalosign: "the uniting of an actual image and a virtual image to the point where they can no longer be distinguished"; and perhaps most importantly in the case of *Memories of Underdevelopment*, the Recollection Image or Mnemosign: "a virtual image which enters into a relationship with the actual image and extends it" (DELEUZE 1989, 335). While DELEUZE links these definitions specifically to the films of Antonioni and Resnais, it is perhaps just as applicable to the objective representation of Sergio and his ideological and psychological disconnect with the Revolution. In this sense, there are several different senses of "Staying Put" that are able to exist on several levels simultaneously: Sergio's simultaneous resistance to his bourgeois origins, his family *and* the Revolution; the continuing underdevelopment of a country still under the cultural and economic aegis of the Batista legacy; as well as Cuba's resistance to the nuclear power game of the October Missile Crisis by "staying put" as a non-aligned nation in the process of becoming something other than it actually is at that present moment (even at the expense of possible nuclear annihilation).

In the latter sense, the Time-Image has nothing to do with flashback, memory or recollection (the basic trope of Sergio's narration). Instead, it makes visible what is concealed even from recollection:

> What we call temporal structure, or direct time-image, clearly goes beyond the purely empirical succession of time – past-present-future. It is, for example, a co-existence of distinct durations, or levels of duration: a single event can belong to several levels: the sheets of past coexist in a non-chronological order (DELEUZE 1989, xii).

Essentially, the cinematic image as we find it in the relationship between fictional and non-fictional space in *Memories of Underdevelopment* cannot be fully integrated into a totality. Instead, it is connected through 'irrational cuts' between non-linked elements so that a confrontation takes place between 'outside' and 'inside'. For DELEUZE (1989, 1), it's from this confrontation and resulting upheaval that thought emerges:

> Is it not rather at the level of the 'mental', in terms of thought? If all the movement-images, perceptions, actions and affects underwent such an upheaval, was this not first of all because a new element burst on to the scene which was to prevent perception being extended into action in order to put it in contact with thought, and, gradually, was to subordinate the image to the demands of new signs which would take it beyond movement?

DELEUZE thus sees modern cinema as constituting a thought outside itself and, concomitantly, the unthought within thought. This is achieved through the "cinema brain" of re-linkages and irrational cuts. This constructive pragmatism represents a pure philosophy of immanence.

This by-play between the Action-Image and Time-Image, between "True" and "False" movement, has led several scholars, not least Julianne BURTON (1977b, 20), to correctly note that ALEA's film is constructed in the form of a cinematographic collage, "not only in its variety, scope, eclectic technique and juxtaposition of evidence from real life on the film canvas, but also in its effect, in the way the combination of the fictional and the documentary, the 'artifice' and the 'reality' exceeds the sum of both parts." In this respect, Brian HENDERSON's (1976, 426–7) seminal distinction between montage and collage is particularly appropriate:

> Montage fragments reality in order to reconstitute it in highly organized, synthetic emotional and intellectual patterns. Collage does not do this; it collects or sticks its fragments together in a way that does not entirely overcome their fragmentation. It seeks to recover its fragments *as fragments*. In regard to overall form, it seeks to bring out the internal relations of its pieces, whereas montage imposes a set of relations upon them and indeed collects or creates its pieces to fill out a pre-existent plan.

More importantly, the use of the Time-Image vs. Action-Image dichotomy pries open the filmic montage and encourages a shift from the hegemony of the edit to the greater, immanent possibilities of the interstice. In *What Is Philosophy?*, DELEUZE and GUATTARI (1994, 164) draw a clear distinction between percepts and perceptions: "Percepts are no longer perceptions; they are independent of a state of those who experience them. Affects are no longer feelings or affections; they go beyond the strength of those who undergo them. Sensations, percepts and affects are *beings* whose validity lies in themselves and exceeds any lived." In short, they exist in the absence of man or woman as a controlling mechanism as they are themselves compounds of percepts and affects: "The work of art is a being of sensation and nothing else: it exists in itself" (DELEUZE and GUATTARI 1994, 164), composed of independent elements such as color, line, shadow and light. Thus Sergio may metaphorically "die" of impotence and inertia at the end of *Memories of Underdevelopment* because he has finally exhausted his sensory-motor connection to the movement-image of the island's collective mobilization, but he lives on in the work of filmic art that gives birth to itself as a being of pure sensation, exceeding lived experience. Again, as DELEUZE and GUATTARI (1994, 173) point out, "The affect goes beyond affections no less than the percept goes beyond perceptions. The affect is not the passage from one lived state to another but man's nonhuman becoming."

The vehicle for this genesis is not the combination of edit and interval that defines the Action-Image – the built in delay that allows the action hero to bide his time within the continuum of the sensory-motor regime, ultimately using the delay to his advantage – but the interstice, the false-movement or fissure through which the Time-Image becomes manifest. As Tom CONLEY (2000, 320) explains:

> The interstice is the interval turned into something infraliminary in a continuum in which an event can no longer be awarded the stability of a 'place' in the space of the image. The interstice becomes what exhausts – and thereby creates – whatever space remains of the image in the sensory-motor tradition. It supersedes the interval and, by doing so, multiplies the happenings of events.

As in Vertov or Godard, in *Memories of Underdevelopment* one image is given, its successor must then be chosen to produce an interstice between the two, as in the formula One + One produces One + One. As CONLEY again notes, it is not a process of association but of differentiation. A potential is given, another must be chosen so that a third, new image is produced as a creative *event*. The advent of the time-image is this very proliferation of events, so that the world is now reinvented in each shot/instance via the Event of the interstice. In *Memories of Underdevelopment*, the interval between the causal montage of the subjective sequences – in which Sergio ostensibly narrates his own story – and the false movement of the interjected documentary sequences (where Sergio is the product of a collective historical moment much larger than himself) proves to be unbridgeable. The edit or splice thus cedes its narrative power to the interstice as the incommensurability of immanent time itself. It is in this "Staying Put" (as a form of Becoming) of immanent time that the body politic, the "People yet to come" – i.e. that which is concealed from thought – is finally made manifest.

ALEA achieves this creation of collective life from the unthought through a method of sequential repetition, where earlier scenes involving Sergio or establishing shots of Havana are repeated later in the film from a different perspective in order to lend new meaning to both the current and earlier incarnation of the sequence. There are two main objectives of this strategy. Firstly, ALEA aims to shift audience sympathy away from Sergio as a character so that they realize, as the director noted above, that they have been caught in an affective trap and are forced to question the reasons for their identification. For example, the early airport sequence where Sergio bids farewell to Laura and his parents is repeated immediately after in flashback as Sergio returns to the city on the bus. As ALEA (1990, 205–6) explains:

> During these scenes, in which no one speaks and all that is shown is the moment of departure, we are constantly observing Sergio, and cannot help but see his ill-concealed mixture of relief

and discomfort. When Sergio returns to the city in the bus and thinks about his relatives who have left, above all his wife, the same scenes are repeated, but now from Sergio's point of view. Only then do we see his wife's and parents' faces; only then do we listen to Sergio's cold, almost cynical voice, which contrasts with the pathetic image of his relatives. The use of a telephoto lens for these images contributes to the isolation from the general ambience of the airport in the faces conjured up by Sergio and helps us understand these images as dreamed or remembered, rather than perceived directly. You could say that first we observed the scene of departure and its effect *upon* Sergio 'objectively,' and then we see it 'subjectively' *from* Sergio's point of view.

The first depiction of the departure thus reveals far less of Sergio's emotional distance than the second, so that evocation is in this case used to undermine the more affirmative viewpoint of objective representation.

The reverse is true in two sequences which depict Sergio tape recording an argument with Laura in their bedroom which tells us a little about the history of their eventual separation. For ALEA (1990, 206),

> it is a silly, frivolous discussion, which begins as a small provocation on his part and proceeds to assume an ever more aggressive tone. The scene accompanying the tape recording shows Sergio alone in their bedroom, amidst the chaos of departure, playing with his wife's jewelry, prolonging the mockery, until little by little the game turns into a bitter corroboration of his cynicism and loneliness.

The first replay is thus presented as a playful, albeit cruel, piece of theater. Sergio drapes Laura's gown over his shoulders and goes through her dressing table drawers. He tells her on the tape that it's being recorded and will be great fun to listen to later (which of course it is). She screams: "You are a monster!" just as ALEA's camera shows Sergio with one of her stockings over his head – brutal and deformed like a bank robber – so that a signifier of her bourgeois beauty becomes a reversed sign of his brutality.

The scene is repeated later in the film, embedded immediately after the seduction of Elena (which includes a scene where Sergio gives his new conquest items from Laura's wardrobe, as if to re-make her in his wife's image) but we now go far beyond the original tape recording and witness the end result of the confrontation. As ALEA (1990, 206–7) points out:

> The earlier scene is repeated and continues past the point where Sergio previously turned off the tape recorder. Sound track and image now go together as Laura falls to the floor in the course of a violent struggle, and then gets up sobbing, insulting Sergio and affirming her decision to leave the country. Once again the film presents first an *evocation* of the action relating to Sergio's frame of mind, and later presents it again, but this time as a *representation* with the character of 'objective' information. (No matter that the second time too the point of departure is a certain state of mind on Sergio's part, nor that Laura's image is seen through Sergio's eyes: the action is still presented with a certain degree of objectivity).

The reprise is far more violent than the initial presentation, showing Laura as she backs away from Sergio in fear and rage: "I never want to see you again!! I'm leaving! I'm going alone. I don't want you to come with me…I won't be a guinea pig for your whims and little games. I'm going to live my own life!" As Julianne BURTON (1977b, 21) notes, this new version of the scene gives Laura a suggestion of autonomy denied to the other women in Sergio's life, while also suggesting that they are all his 'guinea pigs': passive creatures to be shaped and experimented on. More importantly,

> [t]his 'double take' technique requires the viewer's intellectual collaboration and provides a means of increased comprehension within the scope of the film. Point of view is shown to be a crucial determinant of any version of reality. Where the purely subjective novelistic technique only elicits vague suspicions about the narrator's unreliability, impossible to verify within the context of the work, the camera provides an alternate perspective and an external vantage point, whether juxtaposing one person's subjective view to another's or juxtaposing the social and historical context to the individual one.

The second objective of ALEA's repeat strategy is to draw attention away from the evocative, subjective, Sergio-oriented thread of the story to the more objective, collective narrative in which a post-underdevelopment people is in the process of being born. It is in these sequences – through the power of the interstice – that a direct Time-Image comes to override the conventional Action-Image. This is most evident in the opening credits scene of the film, which depicts, *cinema vérité*-style, a nighttime carnival with Cubans of all races dancing to the fiery music of Pello el Afrokán's orchestra. Suddenly gunshots are heard, almost drowned out by the music. A man shoves his way through the dancers. No one stops dancing, even when the revelers see the blood-stained body of a man lying in their midst. The body is then lifted by policemen and carried away through the dancing crowd. According to Paul SCHROEDER (2002, 27), "The setting used here was virtually de rigeur in films made before the Revolution: exotic, sexually charged scenes where violence is quickly absorbed and the party goes on." Thus nothing changes: the dance continues and ALEA's camera finally rests on a black woman's defiant face (see figure below). He freeze frames on her piercing gaze, which remains, as he puts it, "fixed with an expression of subterranean violence. All of this is presented to the spectator from the most 'objective' point of view, the most detached and least engaged, that of the wide shot" (ALEA 1990, 205).

The scene appears to be both a celebration of Cuban culture but also an indictment of its underdevelopment, which is presented through the substratum of violence combined with complete lack of affective engagement. Sergio, in comparison, for all of his bourgeois traits, can subsequently only appear as a model of civilized behavior and his critique of underdevelopment therefore becomes all the more relevant and objective. However, it is only later in the film, when the scene repeats after Sergio walks in the streets, alienated against the tide of a May Day demonstration, that we discover that this scene is shown out of chronological order. It is in fact a flash forward to Sergio's first and only "date" with Noemí. The images are almost the same as before, but the sound is completely different. Instead of Afrokán's stirring, pounding rhythms, the flute-based sound is disconnected and dissonant, without a specific geographical center of gravity. It could be read as an objective correlative of a confused and disassociated frame of mind, clearly dislocated from the overall picture of the shocking violence. Moreover, we now see Sergio and Noemí in the middle of the crowd, so that we are more closely associated with our protagonist's consciousness rather than that of the dancers (see figure below). Although the editing suggests that Sergio sees the assassin, ALEA (1990, 205) points out that he remains a passive spectator, a man who sees and comprehends but refuses to act:

> He is there and yet he is not there. That is, he is present but incapable of becoming involved in the general flow of unconcern, relaxation, unburdening joy, and violence. As much as he tries, he cannot submerge himself in 'his' people's tide. This alternate soundtrack therefore expresses the subjective tension of the protagonist and at the same time maintains our distance from the dance; it keeps us, the spectators, from being passively pulled along by the current. It is no longer the same as at the beginning. Now we feel, along with Sergio, the *distance* that separates him from the environment in which he moves, and this induces us to revise our attitude.

Now that we witness the same violence with Sergio as our point of reference, we are forced to evaluate it in a very different way from our earlier reading at the beginning of the film, when we lacked this subsequent information. Also, what we originally read as present tense from an omniscient objective perspective is now, retroactively seen as a Time-Image set in a future perfect tense, namely a case of "what will have been." Now the penetrating gaze of the Afro-Cuban woman which initially addressed the camera, and by extension the spectator, could be re-read as confronting Sergio and his lack of engagement, jolting him (and us) out of our voyeuristic complacency and forcing us to become conscious observers and participants in what is about to unfold, both when we re-view the film on subsequent occasions and also the closing scenes during the Missile Crisis. Thus, argues ALEA (1990, 205), "seeing it for a second time, from another perspective, and relating it to the central character about whom we already have sufficient information to predict his tragic destiny, the metaphor is expanded. It stretches beyond its initial direct and contingent significance; it opens up and leads to considerations about the reality within which the protagonist is trapped and which he is incapable of understanding in any depth."

This is particularly apparent in the film's conclusion, which expands Sergio's relationship to the collective by relating it to the formal properties of the filmic apparatus as a whole. For example, in early sequences, where Corrieri is inserted into hidden camera street scenes, we see the faces of the Cuban people as Sergio sees them: generally sad, exhausted, tired and unhappy. This reading is reinforced by Sergio's voice over: "What meaning does life have for them…? And for me? What meaning does it have for me…? But I'm not like them." "Nevertheless," notes ALEA (1990, 207), "Sergio's face remains frozen on the screen, showing him to be equally unhappy. (Here again we have jumped from the 'subjective' to the 'objective')." Later, during the October Missile Crisis, as the island mobilizes against potential annihilation while Kennedy and Castro propagandize their respective positions on television, we see the faces of the people again,

> but this time they reveal a state of mind in open contrast with Sergio's, who walks the streets preoccupied and afraid of the atomic disaster which threatens everybody and which he seems to feel more than the rest. In both cases the images of faces are 'objective' in that these are real faces filmed at odd moments in the street. Nevertheless, in each case these faces' meaning is quite different. If at first we get a desolate impression, it is because obviously the protagonist, with whom at first we tend to identify, projects his own emotional state onto the reality that surrounds him and brings us to see it through his eyes. This is the reality that he sees, that he chooses, not what is strictly to be called objective reality. Neither do the faces at the end constitute objective reality in and of themselves, but they draw us much closer to it because they negate the previous impression without totally cancelling it out (ALEA 1990, 207).

It's important to note however, that for ALEA (1990, 207) "The truth lies neither in the one nor the other, nor even in the sum of the two, but rather in what the confrontation between the two sequences and the main character suggests to the spectator in the overall context of the film." This is particularly evident when one considers that although Sergio is a bourgeois with little or nothing in common with either the "man in the street" or the truly revolutionary intellectual, he still perceives and says things about his surrounding reality that are often critically pertinent and well-intentioned, observations that in themselves become a catalyst or stimulus for thought, much like DELEUZE's ideal of the cinematic body. This multilateral view of the film's ideological object should not be seen as ambivalence, ambiguity or indeterminacy. Instead, as ALEA (1990, 207) asserts,

> [i]t is rather the expression of contradictions whose purpose in the film is none other than to contribute to the concerns and impulses for action that we wish to awaken in the spectator. It

thus becomes an incentive to stand at a distance from the images, and in this way encourages a critical attitude, that is, a 'choosing of sides.'

However, as we have seen, this distance can also be a catalyst for lack of engagement, as evidenced by the telescope on Sergio's balcony, the symbol of his voyeuristic detachment. As ALEA (1990, 208–9) explains, "The telescope on his balcony is the most direct symbol possible of Sergio's attitude toward reality: he sees everything from above and at a distance; he is capable of judging reality – from the subjective point of view, of course – but he cannot actively participate in it. This character judges everything, including himself, but his judgment is not always lucid even though it is sometimes quite shrewd." Thus, after he returns from leaving Laura at the airport, Sergio observes Havana through the eyepiece of the telescope (see figure above left), (ALEA pans across the rooftops using a masked point-of-view shot throughout). We see a couple kissing as they sit by the rooftop pool of the Hotel Capri (see figure above right), then ships entering the harbor. "Here everything remains the same," notes Sergio. "All of a sudden it looks like a set, a city made of cardboard." We then see children playing in school, the statue of General Antonio Maceo (the black Dominican leader, martyr of wars of independence vs. Spain, 1868–1878), the ruined foundation of the monument to the US battleship, Maine (but missing its imperial eagle). Slowly, we note brief signs that it's not all business as usual in Havana: a freighter in the harbor, followed by the emergence of a small torpedo boat. Finally, a telescope shot of a billboard: ESTA HUMANIDAD HA DICHO BASTA Y HA ECHADO A ANDAR: an excerpt from the last paragraph of the Second Declaration of Havana, proclaimed by Castro on 4 February, 1962. Sergio, as one might expect, reads the statement personally and ironically. "Like my parents, like Laura, and they won't stop until they get to Miami. Yet today everything looks so different. Have I changed, or has the city changed?" The clear answer is that both have changed, but it takes ALEA's film to show us that the city has changed far more than Sergio ever will.

This is borne out by the film's conclusion, which acts as both a repeat and a historical updating of this sequence. Seeking refuge in his apartment in the face of the impending crisis, Sergio aims his telescope at the stars and moon (symbols of romantic escape) instead of the brutal reality of a blockaded Havana preparing for the worst. ALEA uses nervous jump cutting between images of Sergio's fidgety indecision inside the apartment and the collective determination of the armed militia outside as darkness descends on the city. At daybreak, the final shots of the film show the telescope on the balcony but this time without Sergio's presence. Freed from the shackles of its protagonist's evocative point-of-view, ALEA camera is now able to pan objectively (and by extension, collectively) across the city as Havana mobilizes its defences. SCHROEDER (2002, 49) argues that this strategy forces us to ask ourselves, "As I look at the city-scape from Sergio's vantage point, will I remain a spectator like Sergio, or will I become a participant in the evolution from alienated individualism to committed socialism?"

The final shot provides a possible answer: we pan left and right before zooming in on the Malecón as some cars and a convoy of army trucks pulling heavy artillery move along the waterfront. This is a clear visual cue that the only viable answer is to become a participant at the very moment that the people – born out of the spectre of annihilation – truly comes into being. On the other hand, the filmic spectator is given no easy identification because the interstice between evocation and objectification remains. "As the film proceeds," argues ALEA (1990, 210),

> the spectators should become progressively more aware, through the destruction suffered by the protagonist, of their own situation, and the inconsistency of having momentarily identified with Sergio. Thus, when the film is over, the viewers do not come out satisfied. The passions have not been discharged, but quite the contrary: they have been filled with concerns which should first lead to action upon oneself and then put upon the reality in which one lives. It is therefore a question of a revolutionary act: a conscious grasp of one's own contradictions, the impulse to achieve coherence and project oneself actively on reality.

It is in this sense that *Memories of Underdevelopment* assumes many of the characteristics of a minor literature. For DELEUZE, major literatures always maintain a border between the political and the private, setting up a minimum distance or evolution. In contrast, minor literatures make the private affair political – producing a verdict of life or death, survival or destruction. This inevitably leads to a contradiction: the impossibility of escaping from the group and the simultaneous impossibility of being satisfied with it. It is at this point that Third Cinema steps in with a solution. The act of "staying put" produces a constructive memory of underdevelopment that acts as a contact zone between outside and inside, the people's business and private business, the people who are missing and the "I" who is absent. In this sense, cinema brings collective conditions together through itself (via its story-telling function), thus producing collective utterances. In this regard, story telling *is* memory, and this memory is the invention of a people to come. As DELEUZE (1989, 224) puts it:

> As a general rule, third world cinema has this aim: through trance or crisis, to constitute an assemblage which brings real parties together, in order to make them produce collective utterances as the prefiguration of the people who are missing…

REFERENCES

ALEA, T. G. (1985): I Wasn't Always a Filmmaker. Cinéaste 14 (1), 36–8.

ALEA, T. G. (1990): Memories of *Memories*. *Memories of Underdevelopment, Rutgers Films in Print Series, Vol. 15*. New Brunswick & London, 199–211.

ALEXANDER, W. (1985): Jorge Sanjinéz and Tomás Gutierrez Alea: Class, Film Language and Popular Cinema. *Jump Cut* 30, 45–8.

BURTON, J. (1977a): "Individual Fulfillment and Collective Achievement": An Interview with Tomás Gutiérrez Alea. *Cinéaste* 8 (1), 8–15, 59.

BURTON, J. (1977b): Memories of Underdevelopment in the Land of Overdevelopment. *Cinéaste* 8 (1), 16–21.

CHANAN, M. (1985): *The Cuban Image: Cinema and Cultural Politics in Cuba*. London.

CHANAN, M. (1990a): Lessons of Experience. *Memories of Underdevelopment*, Rutgers Films in Print Series, Vol.15. New Brunswick & London, 3–14.

CHANAN, M. (1990b): Tomás Gutiérrez Alea: A Biographical Sketch. *Memories of Underdevelopment, Rutgers Films in Print Series*, Vol. 15. New Brunswick & London, 15–22.

CONLEY, T. (2000): The Film Event: From Interval to Interstice. FLAXMAN, G. (Ed.): *The Brain is the Screen: Deleuze and the Philosophy of Cinema*. Minneapolis, 303–25.

DELEUZE, G. (1986): Cinema 1: *The Movement-Image*. Minneapolis.

DELEUZE, G. (1989): Cinema 2: *The Time Image*. Minneapolis.

DELEUZE, G. and F. GUATTARI (1994): *What Is Philosophy?* New York.

DESNOES, E. (1990): Inconsolable Memories. *Memories of Underdevelopment, Rutgers Films in Print Series, Vol. 15*. New Brunswick & London, 115–75.

FANON, F. (1966): *The Wretched of the Earth*. New York.

FERNANDEZ, E. (1990): "Witnesses Always Everwhere": The Rhetorical Strategies of *Memories of Underdevelopment*. *Memories of Underdevelopment, Rutgers Films in Print Series, Vol. 15*. New Brunswick & London, 248–52.

GABRIEL, T. H. (1982): *Third Cinema in the Third World*. Ann Arbor.

GABRIEL, T. H. (1991): Towards a Critical Theory of Third World Films. PINES, J and P. WILLEMEN (Eds.): *Questions of Third Cinema*. London, 30–52.

HENDERSON, B. (1976): Towards a Non-Bourgeois Camera Style. NICHOLS, B. (Ed.): *Movies and Methods*. Berkeley & Los Angeles, 422–38.

MILLER, P. B. (1999): Memories of Underdevelopment, Thirty Years Later: An Interview with Sergio Corrieri. *Cinéaste* 25 (1), 20–23.

MYERSON, M. (Ed.) (1973): *Memories of Underdevelopment: The Revolutionary Films of Cuba*. New York.

SCHROEDER, P. A. (2002): *Tomás Gutiérrez Alea: The Dialectics of a Filmmaker*. London & New York.

WEST, D. (1999): Memories of Underdevelopment. *Cinéaste*, 25 (1), 21.

Fernando J. Bosco

DISAPPEARING, STRUGGLING AND RESISTING IN BUENOS AIRES: CHRONICLE OF AN ESCAPE

INTRODUCTION: ESCAPING THE DISAPPEARANCES

Crónica de Una Fuga/Buenos Aires 1977 (Chronicle of an Escape) is a recent example of a film genre that seeks to document the abuses committed by the military government of Argentina between 1976 and 1983.[1] The film reconstructs, through a script in part based on a memoir and historical research, and partly fictional, the story of a group of young men who are abducted, imprisoned, and physically and psychologically tortured in an illegal detention center in suburban Buenos Aires during several months in 1977. The "disappearances" experienced by the characters in the film are examples of the strategies of state-sponsored terrorism that the military government of Argentina embraced to intimidate and control the population for almost a decade and that resulted in the death of an estimated 30,000 people. *Crónica de Una Fuga* is a film about surviving such disappearances. It is also an intimate film in which the emotional geographies of four men – the socio-spatial relations they develop in the small confines of the prison/home they are forced to share – speak to the larger issues of urban violence and resistance that were characteristic of Buenos Aires and Argentina during those turbulent years.

Crónica de Una Fuga begins by depicting the violent abductions of the film's main character, Claudio, as he is getting home walking the streets of his neighborhood. Claudio is blindfolded and taken away, without knowing what awaits

---

1 Crónica de Una Fuga was released to wide acclaim in Argentina in 2006 and was also shown internationally. It was nominated for awards both at the Cannes Film Festival and at the Independent Spirits Awards in the United States and won Best Film at the Lima Latin American Film Festival in 2007.

him. He soon finds himself in a secret detention center that has been strategically set up in a large house in a residential neighborhood in suburban Buenos Aires. Claudio's experience is similar to that of the other three main characters in the film, Guillermo, Vasco and Gallego. The three of them are also abducted unexpectedly in their homes or as they go about their daily lives, often in front of powerless family members, friends or neighbors. The film develops a narrative built around these four men's experiences of 120 days of torture, forced labor, and imprisonment in the house. During this time, the men hardly see daylight, the outside world, their own faces or those of their captors. They endure painful physical and psychological torture in confined spaces, but they also develop relationships that allow them to resist and to imagine the possibility of escaping.

When the four men are the only surviving prisoners in the house and death looms near, they execute a daring escape. They jump out of windows and run through the expansive grounds of the house that is serving as their prison. It is midnight, and they find themselves wandering the same streets of suburban Buenos Aires from which they were abducted as their abductors are in close proximity searching for them. They are naked, beaten, dirty and malnourished. They hide overnight in a garage, stealing some clothes and getting help from a fearful neighbor, until they dare come out to the streets in the morning. The film ends with the emotional separation of the characters, each man going his own way. After being disappeared for over four months, they will attempt to find their way home. Escaping the house, however, is not the end of their experiences as disappeared people. Claudio, Guillermo and Gallego will have to leave the country to avoid being captured again and will become spatially dislocated from the world they know and the places they belong. Vasco will disappear and be imprisoned again, this time for a period of five years, only to reappear alive again in 1983 as the military government falls and Argentina returns to democracy.

Crónica de Una Fuga is an intense and intimate film that creatively plays with the powerful figure of the disappeared person to speak to issues of visibility and invisibility in the city and to reflect on the struggles and rights of certain people to be part of the public, or perhaps simply of the right to be. By presenting a story centered on some of the smallest geographical scales such as the body and the home, *Crónica de Una Fuga* illustrates how violence and resistance are woven together in struggles embedded in larger processes of social and political change in which cities and urban settings play a crucial role. In doing so, *Crónica de Una Fuga* transcends its specific temporalities and spatialities and connects to broader themes of film, urbanism, and the right to the city.

A TECHNOLOGY OF MEMORY, BUT MORE THAN A CULT OF GHOSTS

State-sponsored terrorism and the disappearances in Argentina led to the mobilization of a large and effective human rights movement and, in turn, to a large body of scholarly and artistic work on the subject. There is ample academic work that has analyzed the performative and artistic dimensions of the politics of memory and justice that emerged with democratization after the fall of the military government in 1983 (TAYLOR 1997; BOSCO 2004, 2006). Such politics of memory continue to play a critical role in the dynamics of post-conflict Argentine society today.

There have been many influential artistic and media engagements with the relation between human rights abuses and the politics of memory in Argentina, and film occupies one of the most prominent roles. Marita STURKEN (1997) argues that films are powerful technologies of memory: they embody memory and they become embedded in the power dynamics of its production. Films about events of the past play a crucial part in the construction of *collective cultural memory*, or the entanglement of cultural products and cultural historical meaning outside the avenues of formal historical discourse (STURKEN 1997, 3). *Crónica de una Fuga* is one piece of an extensive repertoire of Argentine artistic cinematic production that works to construct a critical collective cultural memory of the subject and that some now identify as constituting its own genre as the "cinema of the dis-appeared".[2] In this cinema, a number of Argentine filmmakers have been experi-menting with "...ways of creating private biography through stories with strong historic and narrative implications...as a form of processing [their] own past" (AMADO 2009, 38). Some film scholars, however, see problems with this particular genre. SOSA (2008), for example, recognizes the value of these films as technologies of memory but also argues that the cinema of the disappeared tends to create

---

2   In that genre, Crónica de una Fuga joins other famous films such as Luis Puenzo's "The Official Story", winner of the Oscar for best foreign film in 1985

a certain kind of narrative, a sort of monumental view of the past that tends to establish a cult of ghosts [where] the unwritten rule...stipulates that descendants as survivors must honor the name of their vanished parents and friends by raising strong statements in relation to justice, truth and the right of identity (SOSA 2008, 251).

While being a film about the disappeared, *Crónica de una Fuga* does not fall in the cult of ghosts trap. Several characteristics make the film different from other ones in the same genre, in particular the way in which the politics of such a recent past come into play.

There is a tradition of dealing with issues of the social present and social change in urban settings in Argentine movies that dates back (at least) to the *Liberación* and "Third Cinema" (DIXON 2008) politically committed documentaries of the 1960s, in which filmmakers attempted to use this medium to increase collective political consciousness. However, as AGUILAR (2008, 34) explains, the main difference between Argentine films and documentaries from the 1960s and recent productions dealing with political struggles and social change such as *Crónica de Una Fuga* is seen in the ways in which "...the urgency of 'the hour' (in the 1960s) is replaced by the retrospective vision of the 'memories' which, with a touch of nostalgia, permeates almost all...productions of recent years."

While *Crónica de Una Fuga* deals with a sensitive topic in recent Argentine history, the narrative mostly centers on the lived experiences of the characters as they spend time in captivity in the house. Except for some references to their activities at the time of their abductions, the viewer does not learn too much about the lives of the characters prior to their disappearances. None of them are presented as heroes and the film does not spend time trying to proselytize to the audience. Neither the leftist, revolutionary tendencies of Guillermo nor the right-wing, racist, homophobic and violent actions of the guards who imprisoned him and his companions are overemphasized. Claudio, the main character, does not make grand political statements, does not want to be a martyr for any cause, and even shows some resentment towards some of the more politically engaged men being held captive together with him. The result is that the film does not provoke much political reflection; rather, it provides an intimate narrative of the affective relationships between men in an extreme situation.

Because of these characteristics, *Crónica de una Fuga* can also be seen as part of a new generation of films belonging to the "New Argentine Cinema" (NAC). *Crónica de una Fuga* was directed and written by Israel Adrian CAETANO, whose work is often mentioned alongside other filmmakers in this movement.[3] Even though the NAC is very diverse, its films often show an inclination to deal with everyday life in urban settings and with the social present or the very recent past. *Crónica de una Fuga* straddles the cinematic spaces that exist between the cinema of the disappeared that characterized the earlier post-dictatorship productions and

---

3   Other filmmakers in this group include Lucrecia Martel, Pablo Trapero, Daniel Burman and Lucia Puenzo.

the most recent NAC of the 21st century, and as such, the film demands more than a straightforward reading.

There is no doubt that *Crónica de Una Fuga* is a film made to document the suffering, resilience and resistance of victims of the last Argentine dictatorship. *Crónica de una Fuga* shows how, by refusing to give up hope for a chance to live free again outside the detention center, some people managed to survive. As a result of their actions, a story about an almost invisible recent past that still haunts Argentine society today came to light. Without taking anything away from the important work that the film does in the construction of collective cultural memory about the recent past in Argentina and without trying to erase the memories and work of the victims who suffered under the Argentine dictatorship, I see additional elements in the film that make it relevant to other spatialities and temporalities. Engaging with *Crónica de una Fuga* only as a technology of memory precludes seeing how the film also speaks to broader issues about the relations between political violence and social control in urban settings. Specifically, I see *Crónica de una Fuga* as a way to relate the particular events in Argentina in the late 1970s to current strategies of neoliberal urban governance that also rely on violent techniques of surveillance to manage urban public space and control populations. Moreover, I argue that the experiences of the characters in *Crónica de Una Fuga* provide an opportunity to explore the film's engagement with the politics of resilience and resistance that can develop in the context of the house and the home in conditions of adversity.

NEOLIBERALISM, URBAN CONTROL, DISAPPEARANCES, BANISHMENT

Contemporary cities under regimes of neoliberal urban governance are dominated by practices that redefine the public sphere in exclusionary ways. In his book "The Right to the City", Don MITCHELL (2003) describes the way in which the criminalization of homelessness in the United States has served to redefine the public sphere so that those with no access to private property have no place to be. HERBERT and BECKETT (2010) have more recently documented how *banishment* (which include a range of social control tools such as exclusions from parks and other open public spaces, criminal trespass, and off-limits orders) has reemerged as a technique of social control targeting disenfranchised people in cities across the world. In the context of the contemporary (global) neoliberal city, banishment tools give the police the power to clear the city of homeless people, of rural and indigenous migrants, including children (SWANSON 2007, 2010) and of day laborers and other migrant workers (CROTTY and BOSCO 2008). Whether it is in cities of the United States or in Latin America, certain categories of people are considered undesirable, dangerous, or simply not wanted in public space or as part of the public sphere because they challenge or threaten the order expected and required by neoliberal urban regimes.

The disappearances that *Crónica de Una Fuga* portrays so vividly predate many of these contemporary techniques of social control and regulation of urban space, but are surprisingly similar to banishment and other techniques of surveillance and control both in their practical elements and their ideological underpinnings. The process of disappearing people who were deemed suspect of opposing the government was part of the Argentine military's attempt to implement a set of interrelated economic, social and political neoliberal reforms at a variety of spatial scales. The plan was consistent with, and similar to, the efforts of other military governments in Latin America at the time, such as the Pinochet regime in neighboring Chile. Naomi KLEIN (2007) argues that horrific violations of human rights such as the ones committed in Argentina were part of a "shock doctrine" that was applied with the deliberate intent of terrorizing the public to prepare the ground for introducing "free-market" neoliberal reforms. As BRENNER and THEODORE (2002, 3) explain, such neoliberal reforms involve, among other things, assaults on organized labor and other political organizations that defend workers, the poor and the marginalized, the shrinking and privatization of public services, the dismantling of welfare programs and the criminalization of the urban poor. All of these were part of the repertoire of adjustments that the military government had planned for Argentina.

Because opposition to such neoliberal reforms was high – in particular in cities such as Buenos Aires – surveillance, intimidation through fear and violence, and tight control of people's movement and daily life in neighborhoods and urban space more generally were key to their successful implementation. With the help of the death squads and secret police and with the collaboration of all the branches of the Argentine armed forces and the federal and provincial police forces, the military government created a network of illegal detention centers in all major urban areas of Argentina and routinely disappeared people from their homes, their places of work and leisure, and from the streets. *Crónica de una Fuga* is a cinematic window into this set of practices of social surveillance and control. What is captivating about the film is its representation, from the perspective of the victims, of the horrific experience of disappearing and of being banished from society, of being denied the right to the city, of being robbed of a place to be. The characters in *Crónica de una Fuga* are considered a threat to the order of the city and the integrity of public space. They are in the public but they are not wanted as part of the public. In the film, Claudio, Guillermo, Vasco and Gallego disappear because they are no longer wanted in the interactions of public life. They are a threat to the military's view of society and to the government's project for a modern (neoliberal, free-market) country.

The film begins with horrific scenes of the process of disappearing people. First, we see members of the secret police violently entering a home and hitting and torturing a woman. Later in the film, we learn that this woman is the mother of Guillermo, who is accused of having leftist political inclinations. After repeated torture and threats of raping her younger daughter, the woman is forced to reveal the whereabouts of her son, who soon after is abducted from his own home. In the next scene, we witness Claudio, the main character and the goalkeeper for a soccer team, being kidnapped in the street during the day as he is walking through his neighborhood after playing a match. He is beaten, and taken into a car as his neighbors and people in the street watch in fear and seem unable to react to what they are witnessing. Claudio asks for an explanation of what is happening to him, and as he is being beaten, he is accused of being a terrorist. We learn that the military was given reliable information by another man who they have also kidnapped. Claudio is accused of having a mimeograph machine that had been used to print pamphlets critical of the military regime. Even though Claudio denies any involvement with terrorism or any kind of political activism (he is indeed, an innocent victim) his captors do not believe him and the process of his disappearance continues to unfold.

Immediately after, we are shown scenes of the car carrying Claudio traveling through a suburban neighborhood and arriving at a large estate. We see a gate that opens towards the grounds and gardens of a very large house that becomes central to the film. A mixture of Victorian and Italianate styles, the house resembles other early 20th century upper-class homes in suburban Buenos Aires, but it stands out in a middle-class neighborhood. This house, however, is no longer a home. Instead, it is a house that hides an illegal detention center, a place nicknamed Atila by the military and secret police. Claudio, already beaten and tortured, is being taken to this place. A door opens, Claudio is pushed inside, and the images of daylight, of the outside of the house and of a neighborhood give way to much darker scenes that become the norm for much of the rest of the film. Claudio has disappeared.

We now see the interior spaces of what is no longer an upper-class family home but rather a horrific prison and torture center. We see empty hallways and closed doors, and we hear screams of pain and suffering. We see stairs leading to even darker spaces, we see chained windows, and we move through the interior spaces of the house, as if trying to slowly uncover a hidden world that is impossible to imagine from the outside. The film changes locations and spatial scales. The outside world, the world of the city and everyday life in neighborhoods, is now secondary and it is not until much later in the movie that the viewer is given a chance to have a peek into this outside world again.

The film becomes claustrophobic. The focus now is on the prisoners' bodies, their torture and suffering, their pain. From now on, a linear narrative begins to unfold that tells the story of the grim 120 days of imprisonment for Claudio and the men he will meet inside this house/prison. As Claudio is taken inside, we discover that he will be joining several other men, including Guillermo, Vasco and Gallego, who have also been illegally imprisoned and who have been living in deplorable conditions for a longer time. They are tied, blindfolded, malnourished, tortured. They scream, they cry, they bleed. The four main characters are joined in the detention center by a rotating cast of other victims who are becoming disappeared from society and who slowly become disappeared/banished from the film as their captors extract whatever information they can and then continue torturing them until they are unusable. They will all be taken back to the outside world again for a very short time, as they are transported to meet the last stage of the disappearance process, their deaths, somewhere else.

This shift in locations and spatial scales occurs again, this time from inside to outside, when Guillermo is taken out of *Atila* for a short period of time, as he is given an opportunity to redeem himself by providing further information about people that the security forces are looking for. Guillermo is blindfolded and put head down in a police car, without knowing where he is going. Whereas before all we saw of Guillermo were close-up shots of his tortured body inside the dark confines of *Atila*, we now see him in an expanded shot, standing and walking in the streets of a neighborhood.

The car stops in the street as people go about their daily life, and Guillermo is let out. His eyes are uncovered as he clumsily walks in the street for the first time in months. He does not know what awaits him but he is forced to walk towards a familiar house, a house he remembers, a place he recognizes. Guillermo wants to run away but he knows there is a gun pointing at him. He is asked to ring a bell and we expect him to find a familiar face when the door opens. Instead, he is allowed in only to quickly find that this house has also been taken over. Rather than a familiar face or a home space, Guillermo finds a secret police intelligence center and is subject to more psychological torture inside another house that is no longer a home. He also learns that his captors know that he has been giving them false information (such as the information that led to Claudio's imprisonment) as a way to save his own friends and relatives. Guillermo was taken outside of *Atila* into what he thought was a familiar home in the world outside his detention center only to find out that, much as himself, the world that he used to know was slowly vanishing, one friend at a time, one house a time, one neighborhood at a time.

Through these sharp contrasts between outside and inside worlds, between public life in the streets of a suburban neighborhood and private, invisible and horrific life in the interior of a makeshift prison, *Crónica de Una Fuga* makes evident the relations between urban surveillance and control under conditions of neoliberalism enforced through violence. Such contrasts invite reflection on the ways in which portrayals of people disappearing from the streets of their own neighborhoods in Argentina in the late 1970s parallel contemporary struggles of social control and regulation of urban space that attempt to clear "undesirable" people from the streets and from public life. In many ways, the events portrayed in *Crónica de Una Fuga* are a window into the urban present through the recent past: they anticipate the consequences of adopting revanchist regimes of urban regulation in support of a broader neoliberal political economy (SMITH 1996). By using violent means to clear the streets of those who are not wanted in the public, such urban regimes attempt to render them invisible, and *Crónica de Una Fuga* shows what becoming invisible entails.

In *Crónica de una Fuga*, the technique of invisibility is the disappearance, which the military perfected over time and which ultimately resulted in the deaths of 30,000 people and in thousands more people leaving the country, hiding in exile in foreign places, and thousands more being robbed of their families, relatives and friends. Many of the components of this technique are not that different from the strategies used to control homeless people today in cities in the United States. For example, MITCHELL (2003) documents how in the early 1990s in Santa Ana, CA, homeless people were picked up from the streets for minor and insignificant crimes and taken to a municipal stadium where they were chained to benches, illegally held as prisoners, and where identification numbers were written on their bodies with indelible ink – presumably so that they could be later identified quickly and similarly processed if found again in public spaces of the city.

By documenting the horrific violence that was characteristic of the process of disappearing people from the streets in Argentina in the 1970s, *Crónica de una Fuga* allows us to understand how, under similar conditions of global neoliberal governance, those who are marginalized from public space and subject to practices of urban control such as banishment today are becoming the "new disappeared". Both the disappearances of the 1970s in Argentina and the contemporary practice of urban banishment have a similar chilling effect. As HERBERT and BECKETT (2010) argue, banishment robs people of places to be. For some of the disappeared in *Crónica de una Fuga*, this also meant robbing them of their lives. For many around the world today, a similar outcome is not unlikely, and it is in that sense that the film transcends temporalities and spatialities to demonstrate the catastrophic effects of the strategies to control public life under neoliberal urban governance regimes.

HOUSE, HOME, AND RESISTANCE

The house where Claudio and the other men are being held captive plays a central role in *Crónica de Una Fuga*. In urban studies, houses and homes also have played a more central role in recent years. Coinciding with the relational turn in conceptualizations of space and place in human geography (MASSEY 2005), houses and homes can be seen as open rather than bounded. HOLLOWAY (2008, 264) argues that this relational view allows us to see how houses and homes influence and are influenced by "...a wider variety of processes urban theorists might more commonly study at broader spatial scales...(allowing us to) produce a more fully urban, urban studies". *Crónica de una Fuga* gets close to this relational view by centering much of the narrative on the scale of the house to illustrate how, through the creation of a home life during the captivity, the characters are capable of developing strategies of resistance that have implications at other spatial scales that transcend the story told in the film.

The house portrayed in the film was known as the *Mansión Seré*, a large estate that was built as the private residence for the *Seré* family in 1864, a wealthy land-owning, rancher family of French origin. The house occupied a 6-hectare site in the city of Castelar, at the time an upper-middle class suburb of single family homes in suburban Buenos Aires. The house was later sold, much of the land surrounding the house was developed, and Castelar became a middle-class suburban neighborhood, a bedroom community in the rapidly growing Buenos Aires metropolitan area. The *Mansión Seré* experienced several transfers of ownership until it came under control of the Argentine Air Force, which operated an air base just a couple of miles away. The Argentine Air Force began using the house as officers' entertainment quarters and, later and secretly, as an illegal detention center from 1977 to 1978, the time frame portrayed in the film. Hundreds of people who disappeared were imprisoned at the *Mansión Sere*. On March 24, 1978, four prisoners were able to escape from the house by tying blankets out of a window on the second floor and evading the guards during the middle of the night. This is the daring escape that *Crónica de Una Fuga* re-creates.

A few days after the daring escape, the house was burned to the ground and then detonated to erase any proof of the material site where the disappearances were taking place. Whatever remained of the house was completely erased from the landscape in 1985, when the city built a soccer field on the site it formerly occupied. The story of the fate of the house appears at the end of the film, with scenes of the military dismantling the illegal prison immediately after the characters escape. Today, the site of the *Mansión Seré* has become a "place of memory" (TILL 2005) in Buenos Aires. In the year 2000, archeological work began at the site to try to uncover the foundations of the house as a way to make visible again that which was lost. The goal was to make the place speak in the present by (re)constructing memories of a painful recent past. As a technology of memory, *Crónica de Una Fuga* can also be seen as part of these attempts to bring back to life the things that have been buried over the years: the narratives and stories from victims and people who were previously detained and tortured at the house. In doing this, both *Crónica de Una Fuga* and the excavations at the site of the former *Mansión Sere* counter and reverse the former military government strategy of disappearances and invisibility.

The film makes these memories visible through a narrative of resistance that centers on daily life for the four characters inside the house. But in *Crónica de Una Fuga*, the house is much more than a dark physical structure. It is, as

MCDOWELL argues (1999, 92), a "site of lived relationships (…) a key link in the relationship between material culture and society." As the film unfolds, we see the four characters build stronger bonds towards each other in the small interior spaces of the house. Claudio, Guillermo, Vasco and Gallego begin to develop friendships that grow stronger even as their bodies weaken and torture continues to haunt them daily.

Such relationships are closely tied to the confines of the house where they are held captive. A trip to the bathroom together is an opportunity to learn about each other. An encounter in a dark attic where they are being tortured is a chance for emotional support. When Guillermo is severely beaten, Vasco tends to him. Together, they figure out how to lower their blindfolds when the guards are not around, discovering their faces as they learn more about each other. As the film progresses they see more and more of each other. They share food. They care for each other. Claudio is stronger, he gives his food to the others and consoles and encourages Gallego, when he is emotionally and psychologically ready to give up and wants to commit suicide. They cry together, but also dream of living again in the outside world together. A division of labor develops in their strategy of resistance. Claudio watches out for guards while Guillermo tries to get hold of a screw that he will later use to unlock a window. Inside the cramped and dark spaces of this house that is their prison, they begin to build a home life and a politics of resilience and resistance.

The ways in which home life in the house/prison are portrayed in *Crónica de Una Fuga* resemble several of the ways in which homes have been characterized in some critical writings in urban studies. In the film, the home is a prison and a site of oppression in much the same way some feminist writers have written about home life for women, in particular as it pertains to the work of social reproduction that women perform in domestic life (HOLLOWAY 2008). In a film where the entire cast is male, patriarchal power relations and gendered divisions of labor develop over time between the captors and the prisoners. As guards spend more time with the prisoners, they begin relying on them for domestic chores. Domestic life at the *Mansion Seré* is oppressive, and Claudio and Gallego are often chosen to take on roles traditionally assigned to women. Sometimes they are allowed to eat some scraps that the guards have left on their plates, but they are still verbally and physically abused.

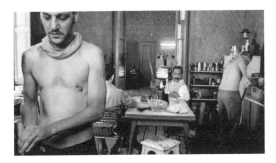

But the gendered division of domestic labor that develops at the *Mansion Seré* over time also provides an opening for the characters to begin planning for an escape and for the film to provide a portrayal of home as a site where politics of resistance can develop. In urban studies, the notion of homeplace as a site of resistance has been put forward more prominently by black feminist scholars. Specifically, Bell HOOKS (1990, 42) argued that one of the ways in which black women resisted oppression was by making homes where they could be affirmed despite poverty, hardship and depravation. In the film, the characters see an opportunity to resist and to plan for an escape precisely as they are forced to do domestic work in the house and as they are emasculated by the guards. The subversion of their domestic slave labor in the home/prison allow Claudio, Guillermo, Vasco and Gallego to perform their identities as subversives, the same identities for which they were disappeared and imprisoned in the first place and that Guillermo embraced (as an activist) but that Claudio denied (as an innocent victim).

Being forced to cook, to clean, to do laundry and to do maintenance work allow Claudio, Guillermo, Vasco and Gallego to access different rooms in the house that they had never seen before. Their activity space inside the house widens while their plans to escape become more solidified. Scenes of small cramped and dark rooms give way to small glimpses of the outside world through a window in the kitchen. A wider scale, new spatialities, and glimpses of hope become apparent. The house is contextualized in place: as Claudio does the dishes, he can see the outside grounds, trees, a dog, a path leading away from the house. The characters develop a new visual perspective and an expanded mental map of the house and the grounds around it. Interestingly, the archaeological work conducted at the site of the *Mansión Seré* since the year 2000 also relied on prisoner survivors' visits to the uncovered foundations of the home and its visible traces on the ground to reconstruct a plan of the house and bring back to life memories and accounts of how they were able to escape. In the film, Claudio, Guillermo, Vasco and Gallego expand their spatial cognition and the house opens up, providing the transition from home as prison to home as a place for resisting, where the characters overcome adversity and develop a politics of subversion that leads to their escape and the end of their disappearances, at least temporarily.

CONCLUSION: EMBODYING URBAN IMAGINARIES

The notion of urban imaginaries, as developed by HUYSSEN (2009), tell us that no city can ever be grasped in its present or past totality by a single perspective or by a single person because cities are palimpsests of diverse experiences, conflicts and memories, constantly changing and continuously produced. HUYSSEN's notion of the urban imaginary joins the ideas of many others geographers and urban theorists who argue that cities are composed of spatial practices that are constantly produced "...by our very visible movements [and by] our recognition and nego-tiation of the built space of our environment" (HUYSSEN 2008, 3). *Crónica de Una Fuga* approximates the concept of the urban imaginary by presenting an embodied and emotional experience of the city and the built environment from the point of view of those who are victims and those who disappear and who are banished. In this sense, the film is unusual because urban scholarly work on the topic often makes us aware of such violent practices from a disembodied and distanced perspective: it is often the scholar-activist who attempts to make us notice the plight of those who are disenfranchised in the city, but the visible bodily expe-rience and suffering of those who are victims is often, paradoxically, rendered invisible.

Media, such as film in general, and *Crónica de Una Fuga* in particular, are a different form of artistic, academic and political expression that make that bodily experience of the urban imaginary visible. By focusing on the emotional geogra-phies that develop at smaller spatial scales – such as the confined cramped spaces of a home and a prison and the changing and deteriorating bodies of the characters – *Crónica de Una Fuga* succeeds in documenting how disappearing and banishment are experienced, and also how surviving and fashioning a politics of resistance out of deplorable circumstances is difficult but sometimes possible. It is in this sense that the film can be seen as making successful spatial and temporal jumps between the disappearances in Argentina in the late 1970s and the strategies of urban control in many cities around the world in the 21st century, including Buenos Aires today.

REFERENCES

AGUILAR, G. (2009): New Argentine Cinema: The People's Prescence. Re-Vista: Harvard Review of Latin America (Fall/Winter Issue): 34–37.

AMADO, A. (2009): Memory, Identity and Film: Blending Past and Present. Re-Vista: Harvard Review of Latin America (Fall/Winter Issue): 38–41.

BELL HOOKS (1990): Homeplace: A Site of Resistance. *Yearning: Race, Gender and Cultural Politics*. Boston, 42.

BOSCO, F. (2004): "Human Rights Politics and Scaled Performances of Memory: Conflicts among the Madres de Plaza de Mayo in Argentina". *Social and Cultural Geography* 5, 3, 381–402.

BOSCO, F. (2006): "The Madres de Plaza de Mayo and Three Decades of Human Rights Activism: Embeddedness, Emotions and Social Movements". *Annals of the Association of American Geographers* 96, 2: 342–365.

BRENNER, N. and N. THEODORE (2002): "Cities and the Geographies of 'Actually Existing Neo-liberalism" *Antipode* 34, 3: 349–379.

CROTTY, S. and F. BOSCO, (2008): "Racial Geographies and the Challenges of Day Labor Formalization: A Case Study from San Diego County". *Journal of Cultural Geography* 25, 3: 223–244.

DIXON, D. (2008): Independent Documentary in the U.S.: The Politics of Personal Passion. LUKINBEAL, C. and S. ZIMMERMANN (Eds.): *The Geography of Cinema-A Cinematic World*. Stuttgart, 65–83.

HERBERT, S. and K. BECKETT (2010) This Home is For Us: Questioning Banishment From the Ground Up. *Social and Cultural Geography* 11, 3: 231–245.

HOLLOWAY, S. (2008): House and Home. HALL, T., P. HUBBARD. and J. R. SHORT (Eds.): *The Sage Companion to the City*. London: Sage, 250–266.

HUYSSEN, A. (2008): Introduction: World Cultures, World Cities. HUYSSEN, A. (Ed.): *Other Cities, Other Worlds: Urban Imaginaries in a Globalizing Age*. Durham and London, 1–23.

KLEIN, N. (2007): *The Shock Doctrine: The Rise of Disaster Capitalism*. New York.

MASSEY, D. (2005): *For Space*. London.

MCDOWELL, L. (1999): *Gender, Identity and Place: Understanding Feminist Geographies*. Minneapolis.

MITCHELL, D. (2003): *The Right to the City: Social Justice and the Fight for Public Space*. New York.

SMITH, N. (1996): *The New Urban Frontier: Gentrification and the Revanchist City*. London.

SOSA, C. (2008): A Counter-narrative of Argentine Mourning: The Headless Woman (2008), directed by Lucrecia Martel. *Theory, Culture and Society* 26, 7–8: 250–262.

STURKEN, M. (1997): *Tangled Memories: The Vietnam War, the Aids Epidemic, and the Politics of Remembering*. Berkely.

SWANSON, K. (2007):. Revanchist urbanism heads south: the regulation of indigenous beggars and street vendors in Ecuador. *Antipode: A Journal of Radical Geography* 39: 708–728.

SWANSON, K. (2010): *Begging as a Path to Progress: Indigenous Women and Children and the Struggle for Ecuador's Urban Spaces*. Geographies of Justice and Social Transformation book series edited by Nik Heynen, Andrew Herod and Melissa W. Wright. Athens.

TAYLOR, D. (1997): *Disappearing Acts: Spectacles of Gender and Nationalism in Argentina's 'Dirty War'*. Durham.

TILL, K. (2005): *The New Berlin: Memory, Politics, Place*. Minneapolis.

.

Pascale Joassart-Marcelli

NEGOTIATING SOCIAL, EMOTIONAL, AND LEGAL BOUNDARIES: UNDOCUMENTED IMMIGRANTS AND THE EVERYDAY FIGHT TO STAY PUT

"After a while you forget…You really feel like you belong."

Mouna, *The Visitor*

"It's very black and white now [after 9/11]: either you belong or you don't."

Peter Shah, immigration lawyer, *The Visitor*

INTRODUCTION

This chapter explores ways in which undocumented immigrants 'fight to stay put' – that is, negotiate and resist the multiple boundaries that preclude belonging and citizenship – through Jean-Pierre and Luc Dardenne's *La Promesse* (1996) and Tom McCarthy's *The Visitor* (2007). In the face of intolerance and cruelty, both films acknowledge the multiple spaces and openings where belonging can occur and illustrate the role of emotions and social interactions in defining these spaces. By revealing that citizenship consists of multiple layers (ethnicity, religion, race, gender, etc.) and intersecting scales (home, neighborhood, city, nation, global communities), the two films discreetly point to a different immigrant politics that moves the struggle beyond traditional rights and duties toward what YOUNG calls "differentiated citizenship" (1999, 264), which acknowledges and respects the different positions, interests and identities of citizens.

In the popular imagination, few people in cities of Western-European and the United States are as vulnerable and disposable as undocumented immigrants from the global south, whose legal status reduces them to temporary workers and transient dwellers – their "illegality" forcing them to live in fear, often finding themselves victims of abuse and exploitation and unable to fight back (CHANCÓN and DAVIS 2006). Many have linked their vulnerability and exploitation to globalizing and gentrifying cities, where undocumented immigrants work as the nannies, janitors, nurses, cooks, busboys, drycleaners and taxi drivers necessary for their very functioning (SASSEN 1998); and yet they must be made invisible to preserve carefully crafted images of global business and entertainment centers. As a result, undocumented immigrants are often confined to "spaces that are structurally as well as geographically, socially, and politically peripheral" (WILLEN 2007, 2).

In academic debates, the socio-spatial exclusion of undocumented immigrants is frequently associated with neoliberalism and the regulation of labor by the state (NEVINS 2002; BAUDER 2006; THEODORE 2007). Strict immigration laws, combined with a lack of enforcement of labor laws, contribute to the creation of low-wage and informal labor markets, keeping other workers in check. Furthermore, these inequities are "managed" and reproduced by a system of governmentality based on racialized constructions of illegal immigrants as lawbreakers, job-stealers, and public burdens (INDA 2006; CHAVEZ 2007; HIEMSTRA 2010).

In these accounts, migrants' illegality is viewed at a macro-scale – the result of global political-economic forces and neoliberal state laws – and the possibilities of belonging are slim if not null. Following GIBSON-GRAHAM (2006), I argue that, while these power relations are real, the "capitalist hegemony" – or the reduction of all social relations to capitalism – prevents us from imagining alternatives and envisioning micro-political ways of performing belonging and fighting exclusion. The relationships which are depicted in the Dardenne brothers' and McCarthy's films reveal interstices, or liminal spaces, where belonging, however briefly, can occur and trigger further change.

In *La Promesse,* 15 year old Igor (Jérémie Reinier) helps his father, Roger (Olivier Gourmet), run an underground construction business in Seraing, a working-class neighborhood of the gray industrial city of Liège, located in the French-speaking part of Belgium. The operation relies on immigrant workers brought over mainly from Eastern Europe and Africa who, in exchange for meager earnings, insalubrious housing and fake residency papers, work on rehabilitating old brick houses for Roger. When one of these workers, Amidou (Rasmane Ouedraogo), is fatally injured while trying to hide from worksite inspectors, Igor promises him he will watch over his wife Assita (Assita Ouedraogo). She had just arrived in Liège a couple of days earlier with her new-born son, after a perilous and expensive underground voyage from Burkina Faso. The film portrays Igor's emerging relationship with Assita and his "coming of age" as he questions the ethic of his father's business.

The Visitor tells a similar story of life-changing relationships that create intimate spaces of understanding, unexpected joy, and belonging. When Walter (Richard Jenkins), a disillusioned university professor from Connecticut, is asked to participate in a conference in New York City, he plans to stay in his Manhattan apartment, which he has not used since his wife's death several years before. Arriving late at night, he is startled to find a young immigrant couple living there. Palestinian-Syrian Tarek (Haaz Sleiman) and Senegalese Zainab (Danai Gurira), both undocumented immigrants, understand quickly that they have been paying rent to a fictitious and criminal landlord. They apologize and decide to leave peacefully in fear of having to deal with the police. Walter is full of remorse and, having found them at the street corner near his apartment, invites them to return until they find another place to stay. In the following days, in a somewhat predictable story line, a friendship develops between Tarek and Walter, influenced by their love of music.

The two films beautifully and unsentimentally depict emerging and trans-forming relationships between people in the midst of adversity, exploitation, xenophobia, and racism. While at the end the lack of legal immigration documents defines the fate of the principal characters in both stories, for ephemeral moments the films give us glimpses of the possibilities of "staying put" by humanizing un-documented immigrants, divulging the roles of place, social relations, every-day practices, and emotional investments in creating a sense of belonging and citizen-ship.

In what follows, I use the experiences of Assita, Amidou, Tarek, and Zainab, as portrayed in *La Promesse* and *The Visitor*, to illustrate these fights to stay put and visit a number of emerging issues in the literature on citizenship and geogra-phies of belonging. After briefly reviewing this recent and growing scholarship, I engage with the films to discuss the role of place by focusing specifically on the home and the city as potential sites of belonging. Second, I emphasize the rela-tional aspect of belonging by investigating how encounters and dialogue challenge and shift boundaries based on racist and sexist notions of who belongs. Third, I address emotional dimensions of citizenship, paying particular attention to the role of objects in informing representations and subjectivities that constitute citizenship.

CITIZENSHIP AND BELONGING

Traditional definitions of citizenship center around the law, either directly as the determinant of one's legal status, or indirectly as the principle underlying the social construction of "illegality." Nation states – and their rigid borders – deter-mine who can become citizens through immigration and naturalization laws. Along with citizenship comes a series of political and social rights, such as the right to vote and eligibility to various social programs. However, to the extent that undocumented immigrants have broken immigration laws, they are often per-ceived as criminals in all domains of life. Reducing or essentializing undocu-mented immigrants to illegal bodies underlies discriminatory discourses and practices limiting immigrant rights in many aspects of their lives. Because they have broken one law, they lose most rights. Thus, within that framework, the fight to stay put begins with challenging immigration laws and restoring rights. Legali-zation or amnesty is seen as a prerequisite to earn the right to live where one wishes.

But "staying put" is more than having a right to stay, it entails a sense of be-longing which rights alone cannot confer, raising questions about the meaning of citizenship and its geographies. Formal notions of citizenship have been criticized on several fronts, particularly for their uncritical focus on the nation state and the law. First, the primacy of the nation state has been challenged by trans-national (LEVITT 2001; BAUBÖCK 2003; EHRKAMP 2005), de-nationalized (APPADURAI 1998), and post-national (SOYSAL 1994; JOPPKE 1999) ideas of citizenship. Increasingly, citizenship has become separated from its ties to the nation state as immigrants keep allegiances to multiple countries (STAEHELI and NAGEL 2006), connect with different communities at various scales (YUVAL-DAVIS 1999; PRATT and YEOH 2003), including local and global communities, and are no longer governed solely by national policies (BRENNER 1999).

Second, the emphasis on the legal framework has been criticized by authors who see citizenship as non-dichotomous and multi-faceted, including legal status, civic participation, cultural incorporation, socio-economic integration and emotional connections. These new approaches emphasize the lived, active and embodied facets of citizenship. In addition, there is growing recognition that the law itself is not static, absolute, and evenly enforced within a nation, but that the legal rights it presumably confers must be chronically (re)negotiated.

Recently, several geographers have challenged traditional concepts of citizenship and focused their attention on "belonging" to a community as a central and inherently spatial component of citizenship (MARTSON and MITCHELL 2004; NAGEL and STAEHELI 2004; LEITNER and EHRKAMP 2006). Borrowing from Lefebvre's idea that the right to the city is defined by *inhabitance* – or living in the city – PURCELL (2003) argues that citizenship ought to be understood as a result of living in and creating urban space. The *perceived* space, *conceived* space, and *lived* space that "*city*zens" produce in their daily routines underlie processes of inclusions and exclusions. Thus, everyday life is critical to understanding citizenship and belonging.

The intimate geographies and mundane practices of everyday are increasingly recognized as important aspects of citizenship (LEY 2004; DESFORGES, JONES and WOODS, 2005; DYCK 2005; DICKINSON, ANDRUCKI, RAWLINS, HALE and COOK 2006), and what takes place in homes, neighborhoods, and communities underlies belonging in ways that reinforce, contradict, challenge, or transform the legal structures set by nation states. Bell HOOKS (2009) writes about making her porch "a small everyday place of antiracist resistance"; a place where she can look at people walking by, greet them, engage in simple gestures of civility, and claim her subjectivity (151–2). For her, belonging is a process of creating places where the rigid boundaries of racism and discrimination can be pushed through intimacies and emotional engagements that generate new understandings.

While legal immigrants, and a fortiori undocumented immigrants, have fewer rights than naturalized and native-born citizens do, they may be animated by a similar desire to belong, to be "home," and thus to become "citizens" in the broad sense of the term. Because belonging is related to social exclusion and inclusion, it is intimately connected to the relationships between migrants and the given host society. These are often viewed through the scholarly lenses of neoliberalism and capitalism, drawing attention to the global forces that govern migration flows and the unequal power relations between migrant labor and domestic capital, which are facilitated by neoliberal state interventions (SAMERS 1999; CHANCÓN and DAVIS 2006; SPARKE 2006; THEODORE 2007). Yet, on an everyday basis, these relations are more complex; they are at the same time accidental and desired, intimate and distant, altruistic and self-centered, tolerant and ignorant. While challenging immigration laws, as well as their interpretation and implementation, is critical to fighting the social exclusion of undocumented immigrants, there are other important – more intimate and mundane – spaces in which the exclusion of immigrants can be resisted and a sense of belonging, albeit partial and temporary, can be experienced.

PLACE AND BELONGING

Citizenship implies a relationship to place – as in belonging to a place or feeling at home. Increasingly, scholars have noted the multiple spaces and scales of migrant allegiance and belonging (PAINTER and PHILO 1995). Feminist geographers have long drawn attention to the spaces of everyday, conceptualizing home and the quotidian as creative events or emergent spaces in which practices of citizenship arise (DYCK 2005; DICKINSON et al. 2008). In questioning the meaning of home for immigrants, consideration must be given to the material and emotional spaces of private dwellings, cities and public spaces, and the transnational ties between communities. Both *La Promesse* and *The Visitor* explore these themes.

At the beginning of each film, a critical scene helps define the importance of home, however humble and precarious, as a place of belonging. In *La Promesse*, we see Igor watching secretly through an opening in the wall as Assita settles in Amidou's dilapidated tenement room after her long voyage. She takes out a small wooden statue and sprinkles a mysterious liquid from a small green bottle throughout the room. Later, Amidou and Assita ceremonially bathe their infant son in the lid of a suitcase, which, as Amidou explains to Igor, is meant to protect him from the bad spirits in his new home. The place is small, cold and dirty, but these performances make it appear warm and secure. Igor, who lives alone with his father in a place not much more salubrious than the dwellings they provide for immigrants, is drawn into this private environment, which appears to be transformed by Assita – reinforcing representations of home as a feminine space.

Figure 1: Assita and Amidou bathing their child

Later, Assita performs other traditional rituals in an effort to establish a home and learn the whereabouts of her husband. She does not know that Amidou has been fatally injured and gruesomely buried underneath the house in cement by Roger who refused to take him to the hospital despite Igor's pleas. Through the eyes of Igor, who has witnessed Amidou's fall and his father's detached and horrific attempt to cover it up, we see both the resilient beauty and the futility of Assita's performance in creating a home at the very place where her husband is buried.

In *The Visitor*, as Walter walks in his Manhattan apartment for the first time in years, the camera rests on a number of objects signaling that the place is now home to someone else. We see a folding table on a cart in the corner, a few white flowers in a cup on the table, a photograph of a couple on a bureau, and a djembe drum. Walter seems to be more disturbed than upset by these visitors, whose in-habiting of the space perhaps remind him of the possibilities that a home brings.

Figure 2: Tarek's drum in Walter's apartment

Together, these scenes illustrate how, in and through simple gestures, immigrants create symbolic and material spaces of belonging in spite of hostile environments. Hanging pictures, cooking meals, taking baths are all parts of the fight to stay put, to create a space where one belongs – defying laws and expectations in minute ways. Neither Amidou and Assista, nor Tarek and Zainab, own the place where they live or even have the right to stay there; they could be thrown in the street any time. Yet through simple performances, they transform these material places into homes. These practices do not lead to an undeniable right to housing or citizenship, but provide temporary asylum which must be rehearsed again and again – often in different physical spaces.

Although the temporary home they create can be a refuge from the hardship of the outside world, in the case of Amidou and Assita it becomes the site of vio-lence and abuse, which remain invisible to everyone but Igor. Roger frequently extorts additional fees from his undocumented tenants for heating fuel, repairs, or visitors. Their exploitation is depicted in a casual, raw and emotionally honest way by the filmmakers. When several East-European tenants become a burden, crowding a room while waiting for an increasingly unlikely trip to the United States, Roger offhandedly disposes of them by reporting them to the authorities. To push Assita to leave the house before she becomes unable to pay rent, he matter-of-factly arranges for his accomplice, Nabile, to rape her in her own room – invading her private space and negating her right to a safe home.

Several scenes in the films link the city, particularly public spaces such as streets and parks, to a sense of belonging, underscoring again the ambivalent role of place. On the one hand, the streets are portrayed as unfriendly and dangerous. After leaving Walter's apartment, Zainab and Tarek are seen standing with boxes and suitcases at a dark and unfriendly street corner – no doubt an attempt to emphasize their vulnerability. A few days later, Zainab fearfully walks away from two policemen too busy watching television through a store window to pay any attention to her, illustrating her almost-constant fear and the emotional violence of possible arrest and deportation. We also learn of Zainab and Tarek's apprehension of the subway, where eventually Tarek gets arrested by two Immigration and Customs Enforcement (ICE) agents dressed in civilian clothes.

Figure 3: Zainab, walking by two policemen

On the other hand, streets and public spaces are also shown as places of tolerance and inclusion. Tarek and Zainab do not live underground. Zainab sells jewelry she makes at a street market where she has several friends and Tarek plays his drum in jazz clubs and public parks. As Walter puts it to the agent who coldly announces to him Tarek's deportation: "he had a life here." In fact, Tarek's connections to the city seem much stronger than Walter's; he walks through the city streets, knows where to get good food, and frequently plays music with drummers from all over the world in Central Park. He dreams to be able to play in the subway one day. By setting fear aside and inhabiting these public spaces, Assita and Tarek resist the reification of streets and parks as dangerous and oppressive, and in the process, contribute to the creation of spaces where difference is valued.

Figure 4: Walter and Tarek, playing drums in New York City's Central Park

The Dardenne brothers are less romantic about the possibility of belonging in the city – owing perhaps to the overtly racist and anti-immigrant attitude found in many economically-struggling European cities. Liège, and particularly the industrial neighborhood of Seraing, lack the glamour and multicultural cosmopolitanism of New York City, and the directors' unpolished documentary-style camera work and location settings highlight the city's grim homogeneity. While Walter's neighborhood is home to Shawarma restaurants, African jazz clubs, ethnic markets, and an "Islamic Council of America" storefront, Seraing consists of the remains of a once-vibrant steel and coal industry, dilapidated brick-row houses, closed-down factories, and barely surviving convenience stores.

One of the most poignant scenes of *La Promesse* shows Assita on the docks along the river Meuse – rusting factories and abandoned coal heaps in the background – holding her son and two large plastic bags containing all she owns. She has just left Roger's house after Igor bravely reveals to her his father's dishonesty regarding her husband's disappearance. She has no place to go. As she waits for Igor's return on the docks, two middle-age white men urinate on her from a bridge, and then ride their motorbikes towards her, yelling racist insults and shattering the content of her bags on the ground. The African statue that had caught Igor's eye a couple of days earlier in Assita and Amidou's room is broken.

In another scene, that night, she tries to get a ride to a hospital in search of urgent assistance for her son, who has caught a dangerously high fever. She waves her arms and screams for attention on the side of the road, but no one stops. One car slows down, but accelerates again as soon as the passengers see her – the drivers presumably turned-away by her skin-color and desperation.

Figure 5: Assita and her baby, with Igor watching in the back

Yet, for Igor, the streets represent an escape from the crimes his father has casually dragged him into. The only vestige of his childhood seems to be a go-kart which he has been building with two neighborhood friends and occasionally rides through the steep cobble-stoned streets of Seraing. As he speeds through the city, he momentarily forgets about the scams that have prematurely surrounded him and the growing tension between him and his father. Assita does not have such an escape, she does not know the city and only experiences its adversities.

Figure 6: Igor and his friends, riding their go-kart in the city streets

In their explorations of the relationship between place and belonging, both films acknowledge the complexity of defining home and the multiple geographies where it is constituted. In addition to revealing the roles of private dwellings and public spaces, they show us how everyday practices and performances of belonging are also connected to transnational geographies. The complex allegiances of immigrants to multiple places are palpable in both movies where the characters never completely break ties with their home country, nor do they envisage returning. After Tarek is taken in custody, his mother Mouna comes to New York City looking for him. She has not heard from him for days and worries. Her late husband had been jailed for years in Syria for political dissent, and she cannot bear the thought of her son in detention. She tells Walter: "This is just like Syria." To Zainab, she says: "I miss Damascus [...] the smell. But this is my home now. I miss it, but I can't live there." Although she lives in the United States, Syria is part of her, it drifts in and out of her mind, but never leaves her body. For her, as for most immigrants, belonging *here* and committing to the *host* country is not possible without being able to acknowledge and nurture the parts that are still anchored *there* in the *home* country.

Because the migrants portrayed in *La Promesse* are so vulnerable and disconnected from the place where they have landed, there are no real instances when commitment to the *host* community is observed. Paradoxically, it is when Assita relates to her home country that she appears to belong most. When she purifies her room, bathes her child to chase the bad spirits, reads chicken entrails for clues regarding her husband, purchases a sheep at the market to prepare for *la fête du mouton*, and seeks help from an African healer, she affirms her desire to be there. These practices, which seem bizarre and exotic to Igor, assert her agency in making this place her home – not separate from her previous community but embedded in it. Allegiance to her home country, through rituals, does not diminish her commitment to her new home where she had planned to reunite with her husband and raise her son. This illustrates that belonging is not the same as assimilation; it does not imply giving up differences and blending into sameness. To belong, one must be able to enact and perform in ways that reflect and produce emotional attachments. By keeping a sheep in an alley below her room, Assita asserts her belonging. Igor nonchalantly asks about the sheep, but he does not ridicule her or stop her from having one there. Instead he helps her secure the animal with debris and in the process acknowledges Assista's belonging and respects her difference.

ENCOUNTERS IN BELONGING

Places are physical spaces, but they are also networks of social processes, inter-actions, and encounters (MASSEY 2005). The role of place in shaping belonging is related to the type of relations they enable. The developing relationships between Walter and Tarek, and Igor and Assita, are central to the sense of belonging expe-rienced by the main characters of both films, including both migrants and natives. In their engagements with Tarek and Assita, Walter and Igor reveal the humanity of undocumented immigrants and slowly deconstruct the social inventions of ille-gality and otherness. By generating new understandings of immigrants' lives, they show us the possibility of inclusion. Yet, in a rather predictable manner, Walter and Igor are the ones most transformed by these encounters, thereby illustrating the dialectic nature of these relations.

Images and narratives of undocumented immigrants as law-breakers have contributed to their dehumanization and underlie neoliberal immigration and immigrant policies (CHANCÓN and DAVIS 2006; CHAVEZ 2007; THEODORE 2007). Taking apart and challenging these social constructs is necessary to generate the potential for social inclusion and belonging. This is what Walter and Igor do – not as a heroic act but as a slow, messy and partial process made possible by an opening, event, or vulnerability in their lives. By engaging in these relations Tarek and Assita actively participate in this inclusionary project and transform hege-monic and oppressive understandings of immigrants.

Tarek, grateful for Walter's willingness to let him and his girlfriend stay in his apartment for a couple of days, reaches out to him and eventually finds a connec-tion through music. Tarek starts giving Walter lessons on how to play the djembe (African drum), invites him to watch him perform at a local jazz club, and later takes him to a drumming session in Central Park. Tarek's trust in Walter con-tributes to a deeper sense of belonging in both men. In contrast, his girlfriend Zainab is suspicious and visibly uncomfortable around Walter. Many additional boundaries exist between Zainab and Walter, including gender, race, and religion. The circumstances do not allow for these to be contested.

In *La Promesse*, Assita deeply distrusts Roger and Igor. She does not separate Igor from his father, Belgians and white people in general, all of whom she sees as oppressors. In a moment of desperation, as she is unable to get a ride to the hospital to get help for her feverish son, she starts throwing stones at Igor and cries: "you make my child sick…you all want him to die…you, your father, your own kind…all of you white people…" By refusing Igor's help, and viewing him as any other white man, she also defines herself as a victim and erects boundaries that reproduce her lack of belonging. She only begins to accept Igor's help when she understands that he is different than his father. As Igor's relationship with his father deteriorates, his connection with Assita gets stronger. By then it is too late.

Brief, mundane and seemingly meaningless encounters with strangers are also important in creating exclusion/inclusion. Immigrant's subjectivities are in flux, being influenced by various representations of immigrants by the host society. On a sunny morning, a white American woman attempts a small conversation with Zainab, who makes and sells jewelry in a street market where artists from different ethnicities sell their exotic goods. She inquires about her country of origin. When Zainab explains that she is from Senegal, the women excitedly replies "I love Cape Town…I was there last year. It's so beautiful." Despite the woman's apparent eagerness to connect, the brief exchange reveals an unbridgeable gap between the two women as Zainab is thrown into a category with which she does not identify herself. The woman's following comment that the bracelet she had been eying was "kind of expensive," also suggests a devaluation of Zainab's work as *ethnic* and thus cheap.

Misconceptions and stereotypes exist in all directions. When Mouna asks the immigration lawyer in charge of her son's case where he is from, she is expecting a certain answer based on his last name and physical appearance – perhaps in hope of finding a personal connection. When he replies that he was born in Queens, the resulting silence says more than any words.

Social inclusion underlies belonging. Thus progressive forms of citizenship must be premised on the respect and understanding of differences. As the stories depicted in these movies suggest, this is not an abstract ideal but the result of social interactions which generate new understandings of diversity and challenge oppressive legacies of privilege.

OBJECTS, BODIES, AND EMOTIONS

Everyday practices of citizenship and belonging also include sensory encounters and corporeal engagements. Touch, smell, taste and sounds have the potential to transmit affect and produce emotions inherent to a sense of belonging. This is particularly well illustrated in Tom McCarthy's *The Visitors* where music and food play a prominent role in shaping relationships between characters, defining their identities, and influencing their feelings of belonging.

At the time he meets Tarek, Walter is living in the past – holding on to the memory of his late wife through classical music, which she used to play professionally. His unhappiness is symbolized by his failed attempts to learn to play classical piano through a succession of teachers. Exposure to Tarek's music creates a rupture that allows him to move on. As Tarek explains during their first informal lesson, "Walter, I know you are a very smart man, but with the drum you have to remember not to think...thinking screws it up...OK?...Forget classical music – leave it behind." Out of the different rhythm of African music emerges embodied belonging to the present.

According to DELEUZE and GUATTARI (1987, 12), music is comparable to a weed or a rhizome; "it sends out lines of flight, like so many 'transformational multiplicities'"; it provides moments where previously unthought and unknown ways of being become possible. Performing music carries an ability to affect and be affected. In particular, improvisation and non-western rhythms like those artfully played by Tarek, defy ordinary notions of intentions and produce new subjectivities. Through its effect on Walter and Tarek's bodies, music changes them and opens up new possibilities.

While music deterritorializes by taking us elsewhere, refrains serve to construct new territories, defending us against anxieties and fears and creating a sense of belonging. Indeed, music – playing it and listening to it – is instrumental to Tarek's belonging. As he shares his passion with Walter and invites him into his world, Tarek feels a sense of agency and pride. Introducing him to the music of Fela Kuti – the famous Nigerian artist Walter had never heard of – brings such a sense of joy to Tarek. During one of Walter's visit in the detention center, Tarek begins to get anxious and frustrated. He tells Walter that he "just needs some music." Together, they play an impromptu short piece by tapping their chests and the desks that separate them. For that brief moment, music allows them to deal with the unspeakable and postpone the inevitable.

Figure 7: Walter and Tarek, "drumming" at the detention center

When Mouna arrives in New York City from Michigan in search of her son, she learns from Walter that he is detained by the US Immigration and Customs Enforcement Agency. Walter, who has witnessed Tarek's arrest and has naively vowed to help him get out, takes her to the detention center. She knows that she cannot go inside to see him – she does not have legal documents allowing her to do so – but she needs to see the place to feel and imagine his presence. At the same time, the thick windowless walls, the desolate streets and the single sign indicating that the center is run by a private corporation, makes his arrest feel more real and his body more distant despite its physical proximity.

Figure 8: Mouna in front of the detention center

While Walter visits Tarek, Mouna waits at a coffee shop around the corner. Although Walter had described it as "not that good," Mouna finds comfort there, drinking tea and speaking Arabic with the flirtatious Egyptian owner. Food and language are sensory experiences that affect belonging and reflect its transnational and complex nature (VALENTINE and SKELTON 2007).

Signs and objects also have the power to generate emotions such as fear and resentment, which reinforce exclusion and work against belonging. The waiting room in the detention center is decorated with posters encouraging immigrants to "know their rights." It also boasts a mural with a bald eagle flying over the New York City skyline against the backdrop of an American flag. In a post-September 11 context, these images are typically associated with patriotism and love of the country. As HO (2009, 791) argues, immigrants' failure to submit to the dominant culture is often "narrated as their failure to love (or identify with) the nation." For Tarek, who is Palestinian-Syrian, these images are a powerful reminder of his exclusion.

The Dardenne brothers' film, in its caustic and unpolished cinema-verité style, does not linger on sentimental moments. Yet the film is emotionally loaded. Until the very end, neither food nor music is shared between Igor and Assita. The latter is too vulnerable, scared, and deprived to invite anyone into her world, underscoring the near impossibility of belonging under the conditions set by the broader structural forces that are reproduced by Roger's racket and Assita's emotional subjectivities. In her words, "there are bad spirits here...we don't see them, but they can see us..."

Igor himself fits oddly in his father's universe. At fifteen, he is no longer in school and is unable to keep his job as a mechanic apprentice because of the pressure to help his father. Although Roger claims to be doing it for his son, he drags him down into an underground world of petty crime. The few ways he shows affection to his son appear to be misguided; he offers him beers and cigarettes, sets him up to have sex with a friend, takes him out to a karaoke bar, lets him drive his van without a permit, and in his spare time, works on finishing a tattoo on his arm. The beer, cigarettes and tattoos symbolize Roger's control over a dark adult world where Igor does not fully belong and which he eventually rejects.

La Promesse reveals the myriad of affective and emotional obstacles to experiencing a true sense of belonging regardless of immigration status. In contrast, *The Visitors* show us the power of emotions in creating citizenship. As Mouna tells Walter on the night before her return trip to Syria, "After a while, you forget [about having no visa]…you really feel that you belong." Together, the films suggest that citizenship – and the fight to stay put – involves much more than legal rights.

CONCLUSION

Belonging is possible in many different ways. While it has traditionally been associated with formal notions of citizenship which grant individuals the legal right to be there, it is also informal and negotiated through practices, performances, and emotions, which underlie relationships to place. As the two films selected for this paper illustrate, belonging is complicated, awkward, tentative, fragile, and unstable. It is also intensely geographic as it reflects social and emotional connections to multiple and interconnected places that are embedded in larger processes and emphasizes the importance of allowing, understanding and respecting differences.

Therefore, the fight to stay put goes beyond traditional national politics. It requires more than a massive protest to ensure immigrant rights or amnesty. It also demands a new politics of citizenship that focuses on creating and preserving places of belonging that affirm difference and are "open to unassimilated otherness" (YOUNG 1990, 241). Instead of denying difference by positing universal rights and common goals, this new politics would provide spaces where the voices of all urban residents can be spoken and heard. To envision such politics, we must start by unraveling the mundane and the quotidian and highlight ways in which immigrants already inhabit the space of everyday and practice citizenship. Focusing on these everyday practices, as *The Visitor* and *La Promesse* do, help us see differences not as fixed and exclusionary identities, but as shifting affinities that can temporarily and spatially overlap.

REFERENCES

APPADURAI, A. (1998): Full Attachement. *Public Culture* 10, 417–450.
BAUBÖCK, R. (2003): Towards a Theory of Immigrant Transnationalism. *International Migration Review* 37, 700–723.
BAUDER, H. (2006): Labor Movement: How Migration Regulates Labor Markets. New York.
BRENNER, N. (1999): Beyond State-centrism: Space, Territoriality, and Geographical Scale in Globalization Studies. *Theory and Society* 28, 39–78.

CHANCÓN, J. A., and M. DAVIS (2006): *No One is Illegal: Fighting Racism and State Violence on the U.S. Mexico Border*. Chicago.

CHAVEZ, L. R. (2007): The Condition of Illegality. *International Migration* 45, 192–196.

DE GENOVA, N. (2002): Migrant "Illegality" and Deportability in Everyday Life. *Annual Review of Anthropology* 31, 419–447.

DELEUZE, G. and F. GUATTARI (1987): A Thousand Plateaus. Translated by B. Massumi. New York.

DESFORGES, L., R. JONES and M. WOODS (2005): New Geographies of Citizenship. *Citizenship Studies* 9(5), 439–451.

DICKINSON, J., M. J. ANDRUCKI, E. RAWLINS, D. HALE and V. COOK (2006): Introduction: Geographies of Everyday Citizenship. *ACME: An International E-Journal for Critical Geographies* 7(2), 100–112.

DYCK, I. (2005): Feminist Geography, the 'Everyday', and Local-Global Relations: Hidden Spaces of Place Making. *The Canadian Geographer* 49(3), 233–243.

EHRKAMP, P. (2005): Placing Identities: Local Attachments and Transitional Ties of Turkish Immigrants in Germany. *Journal of Ethnic and Migration Studies* 31, 345–364.

FENSTER, T. (2005): Gender and the City: The Different Formations of Belonging. NELSON, L. and J. SEAGER (Eds.): *A Companion to Feminist Geography*. Malden, Massachusetts, 242–257.

GIBSON-GRAHAM, J. K. (1996): *The End of Capitalism (As We Knew It)*. Oxford.

GIBSON-GRAHAM, J. K. (2006): *A Postcapitalist Politics*. Minneapolis.

HIEMSTRA, N. (2010): Immigrant "Illegality" as Neoliberal Governmentality in Leadville, Colorado. *Antipode* 42(1), 74–102.

HO, E. L. (2010): Constituting Citizenship Through the Emotions: Singaporean Transmigrants in London. *Annals of the Association of American Geographers* 99(4), 788–804.

HOOKS, B. (2009): *Belonging: A Culture of Place*. New York.

INDA, J. X. (2006): *Targeting Immigrants: Government, Technology, and Ethics*. Malden.

JOPPKE, C. (1999): How Immigration is Changing Citizenship: A Comparative View. *Ethnic and Racial Studies* 22, 629–652.

La Promesse. DVD. Directed by Jean-Pierre and Luc Dardenne. [United States]: New Yorker Films, 1997 (original film release 1996).

LEITNER, H. and P. EHRKAMP (2006): Transnationalism and Migrants' Imaginings of Citizenship. *Environment and Planning A* 38, 1615–32.

LEVITT, P. (2001): *The Transnational Villagers*. Berkeley.

LEY, D. (2004): Transnational Spaces of and Everyday Life. *Transactions of the Institute of British Geographers* 30, 416–432.

MARTSON, S. and K. MITCHELL (2004): Citizens and the State: Citizenship Formations in Space and Time. BARNETT, C. and M. LOW (Eds.): *Spaces of Democracy: Geographical Perspectives on Citizenship, Democracy and Participation*. London, 93–112.

MASSEY, D. (2004): *For Space*. London.

NAGEL, C. and L. STAEHELI (2004): Citizenship, Identity, and Transnational Migration: Arab Immigrants to the United States. *Space and Polity* 8(1), 2–23.

NEVINS, J. (2002): *Operation Gatekeeper: The Rise of the "Illegal Alien" and the Making of the U.S.-Mexico Boundary*. New York.

PAINTER, J. and C. PHILO (1995): Spaces of Citizenship. *Political Geography* 14(2), 107–120.

PRATT, G. and B. YEOH (2003); Transnational (Counter) Topographies. *Gender, Place and Culture* 10(2), 159–166.

PURCELL, M. (2003): Citizenship and the Right to the Global City: Reimagining the Capitalist World Order. *International Journal of Urban and Regional Research* 27(3), 564–90.

SASSEN, S. (1998): *Globalization and Its Discontents*. New York.

SOYSAL, Y. (1994): *The Limits of Citizenship: Migrants and Postnational Membership in Europe*. Chicago.

SPARKE, M. B. (2006): A Neoliberal Nexus: Economy, Security and the Biopolitics of Citizenship at the Border. *Political Geography* 25, 151–180.

STAEHELI, L. and C. NAGEL (2006): Topographies of Home and Citizenship: Arab-American Activists in the United States. *Environment and Planning A* 38, 1599–1614.

THEODORE, N. (2007): Closed Borders, Open Markets: Immigrant day Laborers' Struggle for Economic Rights. LEITNER, H., J. PECK and E. S. SHEPPARD (Eds.): *Contesting Neoliberalism: Urban Frontiers*. New York, 250–265.

The Visitor. DVD. Directed by Thomas McCarthy. [United States]: Anchor Bay Entertainment, 2008 (original film release 2007).

VALENTINE, G. and T. SKELTON (2007): The Right to Be Heard: Citizenship and Language. *Political Geography* 26(2), 121–140.

WILLEN, S. S. (2007): Exploring "Illegal" and "Irregular" Migrants' Lived Experiences of Law and State Power. *International Migration* 45, 2–7.

YOUNG, I. M. (1990): *Justice and the Politics of Difference*. Princeton.

YOUNG, I. M. (1999): Polity and Group Difference: A Critique of the Ideal of Universal Citizenship. SHAFIR, G. (Ed.): *The Citizenship Debates*. Minneapolis.

YUVAL-DAVIS, N. (1999): The Multi-layered Citizen: Citizenship at the Age of "Glocalization." *International Feminist Journal of Politics* 1, 119–136.

Rachel Goffe / Todd Wolfson

ANNIHILATION OF SPACE BY MEDIA: MAKING CONNECTIONS THROUGH A FRAGMENTED LANDSCAPE

INTRODUCTION

Looking at the unfolding practices of an organization based in Philadelphia, this chapter explores how building a media and communications infrastructure can begin to collapse the material and discursive distance between people while connecting the struggles they face in their everyday lives. The Media Mobilizing Project (MMP) engages poor and working people in telling the untold stories[1] of their lives to each other through web, radio, video and TV platforms as well as in face-to-face meetings and media-making; building an infrastructure of technology and relationships between people and groups; and creating access to media, technology and the means to distribute information and stories. In this work, the fight to stay put is broadly conceived as the rejection of a future for Philadelphia shaped by the needs of capital rather than those of people. Staying put in a place means little if the things that enable the fullness of life there – livelihood, education, healthcare, spaces for recreation, and so forth – are being stripped away. This chapter will follow the role of media in coalescing disparate organizing efforts into a broad-based social movement, tracing the fight through settings ranging from the airport holding lot where flexibilized taxicab drivers encounter each other in the city, to parents in Somerset County, Pennsylvania and Philadelphia – joined in a struggle for Head Start and for their families across a racialized geography of poverty, and finally to communities and picket lines inserted into an unevenly disinvested healthcare system in the city and beyond. Ultimately, these locations are all connected through an intricate geography of struggle.

THE VIRTUAL WORLD

The giant communication media...seem determined to present a virtual world, created in the image of what the globalization process requires. In this sense, the world of contemporary news is a world that exists for the VIPs – the very important people. Their everyday lives are

---

1   This sentence "Movements begin with the telling of untold stories" is used by MMP in describing a vision for media's role in movement building.

what is important.... But common people only appear for a moment – when they kill
someone, or when they die (Prometheus Radio 1997).[2]

In the 2008 (and ongoing) city budget crisis, the mayor presented Philadelphia
with "Tight times, Tough choices," and invited residents to a series of community
forums.[3] Designed by the University of Pennsylvania's Project for Civic Engagement, the forums asked residents – organized in small-group break-out sessions –
to arrive at a consensus about what services we could agree to do without from a
series of pre-selected menus. Candidates for elimination carried points linked to
how close each could bring us to our goal: to cut over a hundred million dollars
from the following year's budget. Some participants lobbied others to reach the
goal quickly by adding big ticket items to our cut-list. Most attempted to be
creative and collaborative in response to the charge of civic duty. Noticeably
absent from the menus were any options to close the budget gap without draining
resources from already stressed lives and public programs; even the revenue-
generation alternatives, hastily added in response to public demand, were regressive. Meanwhile at public hearings, residents of gentrifying neighborhoods
expressed rage that an unpopular property-tax relief measure, instituted to reignite
the city's real estate market, was decidedly *off* the table. The ten year property-tax
abatement on substantial renovations and new construction meant longtime residents paid a higher rate for city services than their more affluent new neighbors,
even as tax liens foreclosures were rising.

In this atmosphere, making demands on the local state and capital is rendered
senseless, sectarian or at best idealist. Neoliberal urban governance supports and
extends an idea – that the needs of residents can only be realized through "economic development," which necessitates disciplining social reproduction to the
vagaries of globalized capital and interurban competition (HARVEY 1989; KATZ
2001). That the city presents itself as setting the stage for growth *and therefore*
residents' well-being relies on the erasure of many truths. The mainstream media
shows the lives of poor and working people most often through moments that are
exceptional even though predictable: tragic, criminalized, or unexpected celebration. Severing life events from the rest of life, from history, context, and from
others who might recognize those histories and contexts in their own indi-
vidualizes and naturalizes the effects of economic and state restructuring.

Fragmentation is not only representational. The history of deindustrialization
and deregulation in Philadelphia entailed the dispersal of sites of production, and
moving towards a low wage service economy has increased the contingency and
vulnerability of social reproduction to a fluid local labor market and broader
market upheavals. In response to the fiscal crises of the 1980s, Philadelphia's
make-over into a "World Class City" – a destination for capital, for desirable

---

2   From a transcript of a videotaped statement by Subcomandante Marcos to the 1997 Freeing
     the Media teach-in in New York City.
3   Forums were called "The City Budget: Tight Times, Tough Choices" (Penn Project for Civic
     Engagement 2010)

high-wage workers, and for tourists (HODOS 2002) – has proceeded in part
through the state's abandonment of public goods and social welfare alongside the
creation of selectively punitive re-regulation and regressive revenue extraction.
The remaking of the city and the deepening vulnerability of social reproduction
are brought about via segmented pathways: a vast array of institutional, governmental and market-driven processes even within a single sector, destabilizing the
ability to identify a rational target for a demand (SITES 2007a). This is particularly
the case as social conditions have led to a growing number of people living in
deep poverty that are both isolated and disengaged from the political process
(GOODE and MASKOVSKY 2001).

At work here are two valences of the politics of representation, one having to
do with the exercise of political power and the other having to do with the
positing of explanations. As discussed below, in the case of striking taxi drivers
these two valences initially appear as the problematic: how does a "union" connect with and unify its constituents – a minority of whom are members, and how
does it destabilize the hegemonic portrayal of their demands? Alternative media
projects have often emphasized one sense of representation over the other[4], but
media as a means of collapsing fragmentation must attend to each whilst recognizing them as inseparable moments. For that reason, we suggest this reframing:
How are concepts such as "taxi worker" posed and realized through the process of
building a media and communications infrastructure? Through the cases of Head
Start as a parenting-community, and a defense of "our healthcare system" that
spans workers and communities, we will also consider how building a media and
communications infrastructure can produce persistent collectives (rather than tactical alliances) that exceed the boundaries of organization, sector, identity and
place.

SCENE: TAXI WORKERS STRIKE

In September 2007, taxi drivers across the city parked their cabs in a 24-hour
strike to protest a new regulation that would require Global Positioning System
(GPS) units[5] to be installed in every taxicab and used as the sole interface with the

---

4   On the one hand projects like Indymedia – to which MMP was initially connected – put forth
     a ideology counter to the dominant one, but this is seen as political in itself, without building a
     collective (see MILIONI 2009; KIDD 2003); and on the other hand are media arms of
     independent political parties where the message of the party precedes and speaks to and for
     the party.
5   The director of the PPA's Taxi and Limousine division said its regulatory mission is to ensure
     "a quality service to the riding public," and to safeguard the local taxi industry's viability so
     that it continues to attract drivers and owners (from an interview in MMP 2009a). GPS units,
     linked with an on-board credit card payment system, are presented as a modernizing
     innovation that would make the city fleet more efficient, convenient and safe. Both devices
     were the subject of numerous strikes in Philadelphia between 2006 and 2009, some of which
     were coordinated with the New York Taxi Workers Alliance.

central dispatcher. The Philadelphia Parking Authority (PPA), the industry regulator, argued that the GPS machines were in the interest of all Philadelphians who rely on taxis whether directly for their transportation needs or through the revenue and jobs generated by the hospitality industry. The message put out by the Taxi Workers' Alliance (TWA) of Pennsylvania[6] – that the units were an invasion of drivers' privacy and would reduce already low wages due to their unreliability – seemed sectarian in comparison. The mainstream media portrayed the regulator as the defender of the public, and drivers as self-interested even as they were portrayed as less than enthusiastic about collective action.[7]

There is nothing novel in this story. Yet another face-off between those who claim stewardship of (what are described as) the universal interests of society, and those who struggle to effectively articulate their grievances against the common sense: "Philadelphia best serves its residents by competing for jobs and economic growth," "Cabbies are crooks," and probably the most incredulous "Anybody in their right mind wants a GPS." Of course if the device reports to your boss, your enthusiasm *understandably* dips. The fact that the PPA is not "boss" means drivers are subject to deepening powers of surveillance and oversight without any corresponding responsibility for their welfare. Since they are classified as independent contractors, according to the Commonwealth of Pennsylvania drivers cannot form a union. As such, the courts are the only means through which to air grievances and there is no employer responsible for fulfilling federal labor mandates. Drivers are therefore without minimum wage, workers' compensation, and unemployment insurance protections. Deregulation has reorganized the industry such that each individual is directly subject to the regulator. However, all

---

6   TWA was renamed the Unified Taxi Workers Alliance after a referendum of drivers voted to dissolve the United Brotherhood of Taxicab Drivers and Owner-Operators, a body that had competed with TWA to represent drivers. For clarity, the organization will be referred to as TWA in this writing.

7   The following quotes are provided to illustrate the battle of ideas over who drivers are and what is at stake in their regulation. During a number of strikes contesting the integrated GPS and credit card system The Philadelphia Inquirer, the city's largest daily broadsheet, published several articles about the strike. These pieces quoted PPA officials saying the units would provide efficiency by ensuring "drivers were taking the quickest route to their destination (CLARK 2006, May 17)," would provide convenience for riders by allowing them to pay with credit or debit cards and "improve safety by enabling cabs to be tracked by satellite" (SCHOGOL 2006, 10 April), and that drivers' resistance to GPS units were a cover for their unwillingness to take "anything other than cash" for fares (SLOBODZIAN 2007, September 2). An Action News TV segment went further and claimed that drivers don't like credit because it forces them to claim that income on their taxes (for an online text version that is a partial transcript of that TV segment, see PRADELLI 2008, April 8). Action News also reported PPA officials as saying that the GPS units would "protect underserved communities, especially poorer areas where some drivers prefer not to work (AP 2006, August 16)." Many of these statements play on a popular belief that taxi drivers are dishonest. Several articles included quotes from drivers who were skeptical about the strike (see for example CLARK 2006, May 17). Mainstream media coverage was not devoid of sympathy, but there was no indication of the history of falling profits in the taxicab sector.

of these individuals do not hold the same end of the stick. Some own cabs and some own medallions; others are renters, sometimes from fellow drivers. When the cost of rental slices deep into the earnings of *some* drivers, it supports the internalization of the idea that each cab is an individual producer.

SPACES OF UNITY

The ubiquitous scene described above – a face-off between taxi drivers and city governance[8] – revolves around two senses of representation, both of which appear in a much discussed passage in MARX's *Eighteenth Brumaire* and form the lever of Gayatri SPIVAK's (1988, 1999) argument regarding whether the subaltern can speak. Writing about the position of the French peasantry in Louis Napoleon's 1851 coup d'etat, MARX (1934) observed that as each peasant-holding was almost self-sufficient, occasions for social interchange and interdependence were limited structurally. This isolation was further exacerbated by a lack of communication. Given that, there was no community from which political representation might develop. Laced with irony he writes: "They cannot represent themselves, they must be represented (124)." In the original German the two appearances of "represent" are wholly different words. In the asymmetry of the sentence lurks a question of mediation on several registers. In the first clause, the word *darstellen* means to represent in the sense "to speak of," synonymous with to show or describe. In the second clause, *vertreten* is the sense of represent which means "to speak for," or to stand in for another (SPIVAK 1999). While the latter clause has to do with the authorization of an agent, the first has to do with a circumstance that enables this substitution. In the passage, the impossibility of representing themselves is not an external condition; the necessity for substitution is rooted in the absence of any collective conceptualization, withheld through spatialized material, discursive, and subject-forming processes. Thus "themselves" is not given, something that might have a voice if only we were to listen, nor is it something to which we must be awakened from false consciousness.

SPIVAK's goal in rereading the passage in the *Eighteenth Brumaire* is to assert the irreducibility of concrete subaltern experience to a radical unified political consciousness (challenging Foucault and Deleuze):

> Full class agency (if there were such a thing) is not an ideological transformation of consciousness on the ground level, a desiring identity of the agents and their interests... It is a contestatory replacement as well as an appropriation (a supplementation) of something that is "artificial" to begin with – "economic conditions of existence that separate their mode of life" (1999, 261, emphasis in the original).

---

8   City governance here refers to the sharing of state power across public and private spheres (HARVEY 1989), but some of the referenced organizations – the PPA, the School Reform Commission – now derive authority from the State, not the city.

Commenting on SPIVAK's provocation, GIDWANI (2008) observes that mediation, obvious in "speaking for," also occurs in "speaking of" since the subject necessarily references a system of signifiers that precede and constitute it (873). But what signifiers? Those that reinforce isolation and our subjection to capital? The gap between representation in one sense and representation in the other is the realm of practice, and practice need not indulge a doctrinal adherence to a unified class subject. This we think is the value of storytelling across boundaries, since stories end up being educative, multivalent and reflexive. "Telling untold stories" to each other challenges the effacement of how lives are materially connected, disrupts the fragmentation of resistance and can form lasting bonds between people who care for each other's lives. We argue here that in the process of building this infrastructure – which is at once an infrastructure of technology and relationships – what is made possible is the positing of metaphor-concepts (to use SPIVAK's language) that exceed the isolation of subjects and the isolation figured in the description of their lives, but that these metaphor-concepts are not closed nor wholly determined. In other words, a constellation of simultaneous "we," each of which is an approximation that spans difference, emerge through practices that breach material and discursive boundaries. Each "we" – some of which are organizational ("taxi-workers"), and some of which are cross-network ("the poor") – are constituted through the making of media and the telling of stories.

The spatiality of urban transportation concentrates atomized contractors in the same locations at peak hours of travel. Nodes read off of intersecting networks – the airport lot, the regional train station, a restaurant with ample off-street parking along a busy downtown street – offer a certain regularity to taxis' day to day routes. At the Philadelphia International Airport, taxis are funneled into a holding lot where they await filtration to airport taxi stands, the outflow dependent on airline passengers' travel needs, the inflow based on the fact that this might be the best fare of the day. Though drivers spend many wageless hours here, required by the industry regulator to wait, there is nowhere to get out of the elements, and a lunch truck provides the only opportunity to grab a snack or rehydrate. The only amenities are two poorly ventilated bathrooms that would not pass a health inspection. Here, a panoply of difference collects from behind the wheel every-day, in a sea of blue, green, white, red and yellow cars, unified only through wearing a PPA medallion. To the extent that this purgatory is a place it is because of the practices and relationships that play out here. A bench hosts constellations of drivers as they turn off engines forgoing gas-guzzling air-conditioning and seek respite from summer sun radiating on asphalt; an ad-hoc game of tennis is played on the adjacent concrete island; rolling out mats creates a space for daily prayers shaded by a skyway, feet and hands washed at sinks in the ill-lit bathroom; the hoods of cars become a table encircled by standing conversations, heated and playful.

Into the dispersed network of the taxicab industry, mutual recognition between drivers exceeds their received description as individual producers and the spatiality of a flexibilized industry. Their incidental contact highlights shared experiences of exploitation – a series of regulatory restructurings in addition to rising operating expenses have progressively winnowed their earnings[9]; they are often vilified by the public; they are often the victims of violent crime, including murder; and given their meager incomes, without workers' compensation or health coverage, they often work long hours even when sick or injured. But this recognition does not in itself effectively "speak of" their concerns. An independent contractor who makes a demand for employee protections, a livable wage, and the right to workplace privacy is up against an entrenched common sense that drivers are well-paid and dishonest, eclipsing the falling rate of return that precipitated deregulation in the 1990s.[10] Internally, a membership organization devoid of any right to collective bargaining must make a case to drivers that they are in fact *workers* who can challenge the structuring of their wage relation by a regulator. The contact that TWA makes in these nodes is in large part the only opportunity they have to present the idea "taxi workers" since there are no other lines of regular communication. During the strike, a short video sparked conversations about the action. The video is not sexy. About 6 minutes in length, it is somewhere between rallying cry and political education, firing off several sets of ideas, alternating with documentation of previous actions. The president of TWA contextualizes the situation of drivers in a historic shift in the urban economy, a shift to a service economy that is producing wealth for owners but not workers. He rebuts the PPA's argument: the GPS system does not reliably connect calls with cabs – the units apparently *don't* know where the taxicabs are, based on tests by one local fleet owner. This leads drivers to center city or the airport, areas where they are more likely to be hailed, thus neglecting the neighborhoods – who are presumably among the "riding public" referenced by the PPA. He ends by announcing the inauguration of an international Taxi Workers Alliance that is "in its infant stages but I believe this thing is really gonna last, it's gonna survive and it's gonna be powerful (MMP 2007a)."

The video was posted on TWA and MMP websites, and sent out in an email blast to members and allies of each organization in the network. Most importantly, without a distribution infrastructure amongst drivers or a venue through which TWA could engage them in constructing an alternate frame for their cause, the video was also brought to the airport holding lot, and projected against the frontward wall inside a rented moving truck while drivers cycled through the darkened makeshift viewing booth to watch, lingering before and after to talk. Here, a video simply amplified existing practices, generating a moment in which a concept about who drivers are and what they are up against could be debated, tried on and reworked. Media in this and the campaigns of other network groups are important catalysts, but they are not the full vision for a media and communications infrastructure, to which we will return.

---

9   Recently, TWA has been able to use data that the PPA collects to document that drivers who are not owner-operators earn an average of $4.17 an hour after expenses, working a 12-hour shift. This data is perhaps one positive outcome of the GPS units, given the gravity of cause that empirical evidence can draw.

10 For some of this history, see Driving the American Dream (MMP 2009a).

CONNECTING STRUGGLES

The isolation within the taxicab industry is one example of current challenges to the emergence of a radically different vision for the city. With the dissolution of Fordist regimes of accumulation and the globalization of capitalist production, spaces of production have become more chaotic and dispersed. The neoliberal state is not just minimalist – if it is that[11]; it is also fractured, splintering sites of possible resistance across public, private and hybrid spheres. Much labor and community organizing mirrors civil or state institutional structures, though a group may pull together a temporary coalition to advocate on behalf of, or in common interest with, a particular issue. There are many contributing factors to this mirroring, including narrowly defined funding streams as social services, advocacy and activism have increasingly collected in the non-profit sector (GILMORE 2007), but a key issue is that the structure and language of public policy and labor regulation have become increasingly bureaucratized, decentralized and segmented.[12] This often requires that activists and organizers become specialists, or as described by a healthcare worker in an interview for a segment in MMPtv's[13] pilot episode, "You need someone to stand and speak for you that knows the technicality (MMP 2010)." A striking nurse on the picket line demanded adequate "shift differentials," which when translated means allowing enough time between shifts to protect lives of patients and workers. TWA has had to imagine the implications of being able to record patterns of movement and non-movement and to articulate the temporal effects of credit on drivers' cash flow.[14] Guarding against the incremental encroachments of capital's logic seems to require us to measure our lives in its currencies and voice our concerns in juridical terms. But while specialization of language and of organization may be necessary, it is also an abstraction that leaves the terrain of struggle unchallenged.

Taxi drivers, public school students, patients, healthcare workers, janitors, security guards and others were all struggling long before the current economic crisis, suffering the effects of 40 years of income stagnation, capital mobility, a shrinking social wage, and an increasingly punitive state. In 2008 it was estimated

---

11 PECK and TICKELL (2002) describe a "roll-out" project of neoliberal state restructuring, which includes a buildup of the security functions of the state. The GPS surveillance of taxi drivers by a state agency after they become contractors through deregulation is an example of how these moments work in tandem.

12 As an example, Jason HACKWORTH (2007, Chapter 9) illustrates segmentation in the restructuring of the public housing sector and the challenge that has posed to public housing activism.

13 This summer MMP launched a twice weekly, hour long TV program on a local public access station. The initial episode explores a healthcare system in crisis for patients and workers alike.

14 Credit card payments, in addition to a percentage deduction by the vendor, place an additional cash flow burden on drivers since they pay taxi rental and other expenses upfront, and must wait to be reimbursed for credit card payments by the PPA.

that 24.3%[15] of Philadelphia's residents were living below the official poverty line, a figure which of course grossly undercounts the size of the population that is in dire financial straits. A report compiled by the Philadelphia Research Initiative (2009), *Philadelphia 2009: the State of the City* uses a higher benchmark[16] to designate the "working poor," who are pushed below the poverty line by normal life cycle events – if a family member falls ill or has their hours cut back at work. This includes an additional 1 in 4 Philadelphians, meaning *almost half* the city's residents are poor. These residents are among the public that the city claims to represent, even as it accedes to and actively pursues policies that deepen the chronic crisis of their lives. On top of that chronic state, the acute crisis of city budget shortfalls and market failures have been managed by squeezing workers' wages, challenging pension and healthcare coverage, and jacking up transit fares, utility costs and regressive taxes. At the same time, concessions that are billed as growing the local economy or attracting revenue (in the form of big business and wealthy residents) have been extended without any evidence of their fiscal efficacy while public goods paid for by taxpayers are on the chopping block despite being essential. Beyond the balance sheet, capitalism's crisis has become the pretext for intensifying its discipline over every aspect of our lives. In a recent struggle over the closure of a community hospital, a resident responded "something is wrong with our leadership…why they spend so much money on making downtown so nice…that's all well and good, but when it comes to the little people, we don't have time for you, we ain't got no money for you, get lost! When it comes to someone getting injured or hurt, it's gonna take half an hour to get to any other hospital and by that time they'll be paralyzed or dead (MMP 2009b)." Indeed, Philadelphia is looking more and more like a place where the poor are free to leave, or free to stay and wither or even die.

In this context, the Media Mobilizing Project was conceived as a framework for coalescing a broad based social movement with the political line "a movement to end poverty led by poor and working people united across color lines." Media, because the founders – longtime Philadelphia activists of different lineages – believed an independent media and communications infrastructure could help to collapse isolation within flexibilized workforces; between neighborhoods organizing against displacement or "revenue-generating" projects like casinos and stadiums; between students at underfunded public schools offered the "choice" of charters; across urban and rural fronts, citizens and the undocumented; and beyond each of these sectors, to allow these disparate groups and identities to conceptualize their struggles as somehow connected to each other (WOLFSON, FUNKE and BERGER, Forthcoming). The social forces to which all these fronts of struggle are subject have various institutional and organizational relationships through which they compete or cooperate towards an effective vision for the city and the region. However, sources of an alternate vision are dispersed across herculean

---

15   American Community Survey
16   $35,000 income per year for a family of four

efforts just to maintain ground within discrete social roles. But, taxi drivers are also residents of gentrifying neighborhoods, parents of public school students, patients of community hospitals and the city's public health centers, and users of public libraries, recreation centers and pools, all of which have been embroiled in austerity measures excused by the current economic crisis. It is not just their full lives that need to be represented, but also the fullness of their vision. Via the MMP network, through deepening and proliferating relationships, an expanding body of stories of lives in the city and beyond, and the means to craft and share those stories, disparate fronts of struggle are beginning to name a collective identity.

Currently the *physical* media infrastructure includes five interlinked blogs (the network home, Youth and Education, Labor, Our City Our Voices – focusing on immigrant rights, and All for the Taking – focusing on community-based struggles), a low power FM radio station, a new twice weekly slot on a citywide public access TV station, an expanding "citizen journalism" training program, and the social and material infrastructure required to make, edit and distribute media across these platforms. A number of sub-groups regularly produce themed media: Labor Justice Radio (mostly members of "new labor" unions reporting on labor struggles), Radio Unidad (Latino Americans and recent immigrants playing non-commercial music and hosting the occasional on-air interview with groups like the Coalition of Immokalee Workers), On Blast! (run by PSU members), MMPtv (MMP members and graduates of MMP's "citizen journalism" classes), and an arts and culture committee that is building a citywide space to dream about the future of the movement through its first juried event – four evenings of live per-formance, video shorts, and visual arts entitled "Whose City?" part of the 2010 Philly Live Arts festival.

Initially, MMP acted as the hub through which a number of "network groups," several base-building groups in the city, were connected. Its advisory board was a quarterly meeting of the leaders of these groups. Though decisions were made by the board, joint work was realized within MMP who also produced media and did the work of nurturing relationships between leadership, and developing communication lines between groups. Media produced at that time was edited audio and video, often compiled from a series of interviews through which the groups refused the received framing of their campaigns. These groups included TWA and representatives from other new labor organizations, the Phila-delphia Student Union (PSU) – a youth-led organization of public high school students who demand a quality public school education, Juntos – an immigrant community center that works for justice in the city, and a number of community organizing groups resisting displacement and supporting community self-determi-nation including Casino Free Philadelphia and the African-American Business and Residents' Association. Outside of the advisory board, MMP began by hosting cross-network gatherings: an annual Community Building Dinner, sum-mertime barbecue/game playing afternoons and a weekend long network communication and leadership school. These events are important face-time where the broader membership of the network can build relationships, study the terrain and narrative of their individual and shared struggles, and get an awe-in-spiring sense of the size, depth and variety of experience and concerns that the network spans. Large-scale face-to-face events are supplemented by smaller more intense ones: immersion trips to places like Baltimore, Detroit, West Virginia, etc. both to experience and document conditions and meet people fighting similar struggles in other places. Between MMP-organized network events, the network materialized in sharing information, mutual support of and jointly organized actions. In speaking in support of each others' actions, it became necessary for students to articulate why they were involved in a taxi-workers' prayer vigil out-side the PPA, or a taxi-worker in students' direct action in front of the School Reform Commission. Since 2009, MMP has acquired a physical space where the network can meet, make media, read and watch media together, study, strategize, recharge, learn, and celebrate together. The assertion here is that these practices can subvert sociospatial processes that dissect lives into discrete social roles, dis-tribute organizations across discrete social struggles, and differentiated categories of people across material and social space. In the cases discussed below, media-making creates new ways of thinking across well-worn boundaries of race, urban and rural divides, and workplace vs. community based politics.

PARENTING IN COMMUNITY

Nationwide, Head Start has provided early childhood education, nutrition and health services for low income families since its founding in 1965. In 2009, there were 929,257 children enrolled in all 50 states. The racial distribution of enrolled children is 39% white, 31% African-American, and 34% Hispanic (HOFFMAN 2010). Early in its history, civil rights leaders fought to ensure that the program was not a service provided to children, but would put parents living in poverty in charge – of the leadership of their family, and of a community of families. This vision endures in the language of the federal Head Start program which requires grantees to have "maximum feasible participation of the poor." While many dif-ferent kinds of agencies are eligible for Head Start funding including school dis-tricts and private organizations, the governance of each local Head Start agency involves a body called a policy council. It is the policy council that reviews agency procedures, and makes decisions about staffing, budgeting and policy de-velopment. This body is required to be more than half parents of currently en-rolled children with the remainder composed of members of the community. During their years of engagement, parents learn and pass on leadership skills and survival strategies that impact their lives long after their children have stopped participating in the program.

Head Start has been under attack through successive federal budget cuts over the past decade and through accusations that it is ineffective and that local agen-cies mishandle funds. In 2007 when Head Start was up for reauthorization, Congress proposed limiting the decision making power of policy council, concentrating it instead in the Board of Directors (CARUSO 2006). *Hope in Hard*

Times (MMP 2007b) is a half-hour video produced through a partnership between MMP and the Pennsylvania Head Start Association (PHSA). Ultimately each member of the US Congress received a DVD copy before they voted on the bill, and PHSA – whose website says the video "should be used as a tool to inspire conversation and action" – credits it with a central role in reauthorization. But it was the process of making the video that brought parents in Philadelphia and Somerset County (Pennsylvania) into conversation, realizing they were having to make some of the same choices – like deciding whether to feed themselves or keep up with utility and housing payments. In the words of one parent in Somerset County, "...I'll tell you what, sitting in this room [where the policy council meets], that is how I've been able to take care of my family the best way – not from the bureaucrats, not from the agencies – just one on one talking like this with different people across, not only across the county but across the state, 'cause we all know what we're going through." In another interview with the same parent she said, "You learn and you get a network of friends and a network of – even pro-fessionals – that you might not know the answer to [a question] but you can always ask them and you can ask them 10 years down the road [laughing] like I've done." Out of the lived experience of policy council, parents talked about Head Start as a community of parents caring for a community of children. This is an experience earned through a history of defending "maximum feasible partici-pation." When a new generation of parents was thrust into the fight, the process of making a video became a vehicle for learning that history and for broadening the scale of a community. A struggle for Head Start became a struggle for other fami-lies like them, in other places and in the future.

Living in a country with a racialized geography of poverty, making a video across urban and rural Pennsylvania also meant seeing similarities between the effects of exploitation in places where those who live in poverty are predomi-nantly people of color and places where they are predominantly white. Building relationships between people in these locations and between the stories that are told about people living in poverty by themselves and by others about them is a necessary task given entrenched divisions in United States history, a task begun by the Poor People's Campaign under the leadership of Martin Luther King Jr. and the Southern Christian Leadership Conference. Together with a national net-work of organizations connected through the Poverty Initiative at Union Theo-logical Seminary, MMP continues to put Philadelphia-based groups in contact with poor-led groups in southern Appalachia and western Pennsylvania. In Phila-delphia, battles over gentrification often fall on color lines. But between Appala-chia and Philadelphia members of the MMP network found the common ground uncanny – land struggles in Appalachia were as much a defense of a way of life as in Philadelphia – to set the terms for land use over the market value of mining, real estate, or gaming capital. Back in Philadelphia, when a proposed casino site moved from Fishtown – a white working class neighborhood, to Chinatown – which is a neighborhood of mostly Asian immigrants and Asian-Americans located within the city's central business district, Casino Free Philadelphia (com-posed of Fishtown residents) and Asian Americans United (based in Chinatown) were able to join forces and declare: "No Casinos in the Heart of Our City." While this alliance may be imminently strategic, it is a living marker of lessons learned away from home, and the value of traveling. "United across color lines" is an incomplete project. It is important to acknowledge the very real and violent sedimentations of difference and the varied expression of race and ethnicity in work vs. community-based struggles in the U.S. even as political projects are built that span these "trenches (SITES 2007b)."

SHIFTING STRUCTURES

Making media in moments of protest and crisis have been opportunities to step back from mainstream discussion and frame new ways to think about ourselves. Speaking of these concepts to each other internal to the movement has become part of constituting them in the world. These concepts include "taxi workers" but also as connected to something beyond that, something that until the last few weeks of 2008 was most palpable in those moments when people from across the network gathered in large groups to celebrate, but is beginning to have an everyday life as it reshapes the organizational structure of the network.

With the city budget crisis that began immediately after the 2008 presidential election, a cross-organizational body emerged that studied together for a year to understand better the global economic crisis and the impact and opportunities it created for the network, the state and capital. This body is neither MMP nor any other organization, but is the seed of a new collective identity, one that has begun to take over the leadership of the network and to show up in the structure of stories told by the network. During the taxi workers' strike the president of TWA, Ron Blount, posed the idea that perhaps it was taxi workers who were the defenders of the riding public. *Say what?*

But with the city government's threat to close all the public pools, libraries, recreation centers, a few fire companies, etc. the network began internally to try on making media that was beyond any one campaign. It started with community groups that were challenging the closure of their community hospital and the nurses who joined that effort, some of whom were also residents of the community. The city budget crisis unfolded alongside the national debate on healthcare reform. For the neighborhoods served by Northeastern Hospital, owned by Temple University Health Systems, the healthcare crisis gained a territorial character. By Temple's own admission, the reason for the closure was the disproportionately higher Medicare, Medicaid and uninsured patients of the community hospital (GEORGE 2009). As healthcare capital moves out of disinvested neighborhoods, travel times to emergency and maternity care for poor and working class residents gets critically longer. Healthcare workers' contracts were soon to expire at Temple's main campus, the hospital that was supposed to pick up the former patients of Northeastern, who had itself picked up the former patients of the last hospital Temple closed a few years before. A project began to unfold that took "our healthcare system" as its subject; a story about the city, and beyond. The Northeastern fight was a way to build relationships with community groups, including ones that had been involved in resisting the closure of their library in the budget gap crisis. It was also a way to get to know the healthcare workers union, the Pennsylvania Association of Staff Nurses and Allied Professionals. The video that started to take shape also reached out to ACT-UP Philadelphia who was fighting defunding, newly unionized retirement homes in the suburbs, and uninsured taxi drivers injured on the job. When the video, which had turned into an hour-long pilot episode of MMPtv, was screened months later, participants in all these struggles came out and celebrated and chanted "Act Up, Fight Back!" together for the first time. This had become a struggle that recognized what was on the line when an already overburdened hospital was to draw even more lives into waiting for care, and workers' renewed contracts were to include a gag clause preventing them from disclosing conditions at the hospital, what was on the line was life itself. Not simply the loss, endangerment, or displacement of life, but whether life is to be remade in the image that capital requires. Alongside this recognition is a political project that anyone can have a stake in.

CONCLUSION: MEDIA AS A SPATIAL PRACTICE?

There is a growing literature considering the spatial constitution of social movements. Responding to the mostly aspatial and taxonomic lineages in the sociological social movement theory (see MCADAM, TARROW and TILLY 2001), recent writing has advanced interdisciplinary approaches that draw from geography a more nuanced approach to space, incorporating specific spatial categories as well as the idea that space is not just a container for social processes, but that social and spatial processes are mutually constituted (COBARRUBIAS and PICKLES 2009). MILLER (2000) lays out an approach that entails a detailed empirical investigation of political mobilization in order to explain how geographically variant histories and conditions impact the "propensities for protest" (63) in different places. MILLER discovers a sort of sweet spot where certain institutions coexist with a sense of place and political opportunities as structured by the local state which together provide a territory-based solidarity mobilized around strong grievances and the appeal of achievable goals. Much of this literature is geared towards assessing post-facto the facilitating or limiting conditions for mobilization. Less common is a perspective that grapples with how a density of relationships and commonality of vision might be fostered (or strengthened). COBARRUBIAS and PICKLES redirect inquiry towards spatial practices of movements themselves, and their use of radical cartographies that attempt to incite new cognitive maps. We agree that writing need not foreclose the futures of projects that seek to disrupt and hack into sociospatial processes that create drag on the emergence of social movements.

Answering the charge *what is to be done,* we ask: under present conditions how might media and communications be a means to short-circuit and supplant fragmentation? In MMP's work, this question has emerged as twofold – making connections between groups mobilized around issues understood as separate, and how these issues can instead be understood as connected. While it is useful to talk about each of these moments in turn, in practice they are entwined in "the telling of untold stories." This phrase takes media neither as a product, adequate in itself, nor as subordinate to the movement. Instead, media is seen as a dialogic form, a process through which organizational, sectoral or social collectives are constituted. Communications is a necessary supplement to these stories. The day to day sharing of information between organizations and individuals form the basis for lasting relationships, a shared leadership and a shared strategy out of which grounded institutions of representation can emerge, institutions that can upset capitalism's subordination of life to its role as the source of value.

ACKNOWLEDGMENTS

This chapter draws on the brilliance of many people in the network whose thinking over the years bubbles up in ways we can often no longer distinguish. Among them are Phil Wider, Nijmie Dzurinko, Shivaani Selvaraj, Chris Caruso, Mica Root, Ron Blount, and Desi Burnette. The list of those involved in the project and the media cited is too long to include here. We would also like to thank Christian Anderson for his insightful comments on an earlier draft.

REFERENCES

Associated Press (2006, August 16): Judge Rules Against Cabbies; Another Strike Threatened. Philadelphia, PA: *6abc.com*. Retrieved from: http://abclocal.go.com/wpvi/story?section= resources / traffic&id=4466395.

CARUSO, C. (2006): Turning the Tide of Neoliberalism: A Briefing Paper on the Defense of Head Start. *Unpublished Manuscript prepared for the Poverty Initiative.*

CLARK, V. and J. S. WRITERS (2006, May 17): Taxi strike causes daylong difficulties for city travelers. *The Philadelphia Inquirer,* (pp. B03).

COBARRUBIAS, S. and J. PICKLES. (2009): Spacing Movements: The turn to Cartographies and Mapping Practices. WARF, B. and S. ARIAS (Eds.): *Contemporary Social Movements. The spatial turn: interdisciplinary perspectives.* (pp. 36–58). London.

GEORGE, J. (2009): Temple Ending Inpatient services at Northeastern Hospital. *Philadelphia Business Journal,* online version. Retrieved from: http://www.bizjournals.com/philadelphia/ stories/2009/03/23/daily10.html.

GOODE, J. and J. MASKOVSKY (2001): Afterword. GOODE, J. and J. MASKOVSKY (Eds.): *The New Poverty Studies: The Ethnography of Power, Politics and Impoverished People in the United States.* New York: New York.

GIDWANI, V. (2008): Capitalism's Anxious Whole: Fear, Capture and Escape in the Grundrisse. *Antipode.* 40 (5): 857–878.

GILMORE, R. W. (2007): In the Shadow of the Shadow State. Incite! Women of Color Against Violence (Eds.): *The Revolution will not be Funded: Beyond the Non-Profit Industrial Complex.* (pp. 41–52). Cambridge.

HACKWORTH, J. R. (2007): *The Neoliberal City: Governance, Ideology, and Development in American Urbanism.* Ithaca.

HARVEY, D. (1989): From Managerialism to Entrepreneurialism: The Transformation in Urban Governance in Late Capitalism. *Geografiska Annaler. Series B. Human Geography,* 71, 3–17.

HODOS, J. (2002): Globalization, Regionalism, and Urban Restructuring: The Case of Philadelphia. *Urban Affairs Review,* 37, 3. 358–379.

HOFFMAN, E. (2010): Head Start Participants, Programs, Families and Staff in 2009. (a publication of the Center for Law and Social Policy) *clasp.org.* Retrieved from: http://www.clasp.org/admin/site/publications/files/hs-preschool-pir-2009.pdf.

KATZ, C. (2001): Vagabond Capitalism and the Necessity of Social Reproduction. *Antipode.* 33, 709–728.

KIDD, D. (2003): Indymedia.org: A new communication commons. MCCAUGHEY, M. and M. AYERS (Eds.): *Cyberactivism: Online activism in theory and practice.* (pp. 47–69) New York.

MARX, K. (1934 [1852]): *The Eighteenth Brumaire of Louis Bonaparte.* Moscow.

MCADAM, D., S. G. TARROW and C. TILLY (2001): *Dynamics of contention. Cambridge studies in contentious politics.* Cambridge.

MILLER, B. A. (2000): Geography and Social Movements: Comparing Antinuclear Activism in the Boston Area. *Social Movements, Protest, and Contention, v. 12.* Minneapolis.

MILIONI, D. L. (2009): Probing the online counterpublic sphere: the case of Indymedia Athens. *Media, Culture Society.* 31: 409–431.

MMP (2007a): *Untitled: Taxi Workers Alliance of Pennsylvania's Public Service Announcement.* Online audio. Retrieved from: http://www.phillyimc.org/media/2007/08//41794.mov.

MMP (2007b): *Hope in Hard Times.* Online video. Retrieved from: http://www.paheadstart.org/index.cfm?mm_id=15.

MMP (2009a): *Driving the American Dream.* FILLIPPONE, C. and C. ROGY (Dirs.) Online video. Retrieved from: http://blip.tv/file/2749907/.

MMP (2009b): *Local to Federal: The Struggle for Healthcare.* K. CAROLAN (Prod.) Online audio. Retrieved from: http://allforthetaking.org/listen-human-right-healthcare-northeastern-hospital-ground-zero.

MMP (2010): *Infection in Our Healthcare System.* Online video. Retrieved from: http://mediamobilizing.org/ mmptv-now-live-online-infection-our-healthcare-system.

PECK, J. and A. TICKELL (2002): Neoliberalizing Space. *Antipode.* 34, 380–404.

Penn Project for Civic Engagement 2010: Tight Times, Tough Choices. Retrieved from: http://www.gse.upenn.edu/node/690.

Philadelphia Research Initiative. (2009): *Philadelphia 2009: The State of the City.* Philadelphia: Pew Charitable Trusts.

PRADELLI, C. (2008, April 8): Cab Drivers and Credit Cards. Philadelphia, PA: *6abc.com.* Retrieved from: http://abclocal.go.com/wpvi/story?section=news/local&id=6068406.

Prometheus Radio (1997): *Marcos on Media.* Translated transcript of videotaped statement. Retrieved from: http://www.prometheusradio.org/marcos_on_media.

SCHOGOL, M. and K. R. WRITERS (2006, April 10): Phila. cabdrivers plan protest today; one group urged a strike over new and pending rules. Another reportedly refused to participate. *The Philadelphia Inquirer,* pp. B03.

SITES, W. (2007a): Contesting the Neoliberal City? Theories of Neoliberalism and Urban Strategies of Contention. LEITNER, H., J. PECK and E. S. SHEPPARD (Eds.): *Contesting neoliberalism: urban frontiers.* New York, Guilford Press. 116–138.

SITES, W. (2007b): Beyond trenches and grassroots? Reflections on urban mobilization, fragmentation, and the anti-Wal-Mart campaign in Chicago. *Environment & Planning A.* 39, 2632–2651.

SLOBODZIAN, J. A. and I. S. WRITER (2007, September 2): City cabbies say technology is too taxing. *The Philadelphia Inquirer,* pp. B01.

SPIVAK, G. C. (1999): "History" in *A critique of postcolonial reason: toward a history of the vanishing present.* (pp. 198–311). Cambridge.

WOLFSON, T., P. FUNKE and D. BERGER (Forthcoming): Communications Networks, Movements and the Neoliberal City: The Media Mobilizing Project in Philadelphia. *Transforming Anthropology.*

Amanda Huron

CLAIMING SPACE IN THE AIR AND ON THE BLOCK: THE GEOGRAPHY OF MICRORADIO AND STRUGGLES AGAINST DISPLACEMENT

INTRODUCTION

A radio wave appears to be fleeting. It cannot be seen or touched, apparently ungrounded, an ethereal presence detached from the earth. Yet radio in its smallest forms can be deeply connected to the land. The particular geography of microradio can be a powerful tool for fighting for the right to be in a certain place: the right to stay put over time, to create culture, to dwell. Here, I examine the case of one contemporary microradio station in its struggles against neighborhood displacement, and consider the possibilities for the future.

DISPLACEMENT, DWELLING AND THE RIGHT TO STAY PUT

The thorny question of gentrification-induced displacement first entered the U.S. public sphere in the mid-1970s. Across the country, housing costs were far outstripping incomes: the median cost of a new home doubled between 1970 and 1976, while the median family income rose by only 47% (FRIEDEN and SOLOMON 1977). Ominous headlines like "Housing Costs Outrun Income of Blacks" and "Middle Class Return Displaces Some Urban Poor" appeared regularly in major newspapers like *The Washington Post* and *The New York Times* (FEINBERG 1975; REINHOLD 1977). While displacement due to direct government action was an old story, manifested most recently in federal urban renewal projects (see THURSZ 1966), displacement due to what was sometimes called "private urban renewal" was a new phenomenon (ZEITZ 1979). In 1977 displacement was of enough concern to warrant Congressional hearings, and in 1978 Congress required the U.S. Department of Housing and Urban Development (HUD) to conduct a study on the nature and extent of the phenomenon (HUD 1979; GOLDFIELD 1980).

HUD released its study, *Displacement Report*, in February 1979. The report, which was a survey of the existing research on displacement in American cities, concluded that no one had been able to prove that displacement was a statistically significant problem. The authors cite, for example, an 18-city survey which found that fewer than 100–200 households were displaced in each city annually. Another study cited estimated that 500,000 U.S. households were displaced each year between 1974 and 1976 – just 3.8% of the people who had moved in that time.

The available data, the authors argued, simply did not back up the media hype over displacement. "The major conclusion from this survey of displacement studies," the authors wrote, "is that very little reliable information exists" (HUD 1979, 30). The stories about displacement were, a HUD official wrote later that year, simply that: stories. Until further research showed displacement was a problem, the federal government was not going to address it (SUMKA 1979).

Statistics told one story. But low-income people in central neighborhoods in many American cities seemed to be experiencing something else. At a national meeting in 1977, Legal Services attorneys, who worked with low-income people on a daily basis, identified displacement as their top priority for "national level research and coordinated action," and formed the Legal Services Anti-Displacement Task Force to tackle the issue (HARTMAN, KEATING et al. 1982, 1). In 1981, two members of the task force, Richard LEGATES and Chester HARTMAN, published a critique of the HUD displacement study, meticulously taking apart its arguments and challenging both its statistics and its philosophy. Displacement, they argued, was in fact a major problem, both in terms of the numbers of people affected and the personal impacts of forced moves (LEGATES and HARTMAN 1981). The following year the National Housing Law Project published the book *Displacement: How to Fight It*, in which members of the task force, including HARTMAN and LEGATES, outlined practical methods for fighting displacement. It is here that the authors put forth the idea of a "right to stay put," writing:

> We put forward the twin goals of absolute defense of 'the right to stay put' and absolute requirement that there be one-for-one (or more) replacement of lower-rent units withdrawn from the market, realizing that many local groups will not be able to win such demands. For them, the anti-displacement battle may turn into a fight to get all they can to compensate for the pains and costs of dislocation, or to enable some residents to remain. But the push towards basic goals is always important (HARTMAN, KEATING et al. 1982, 5).

HARTMAN elaborates on this "basic goal" in his 1984 article, "The Right to Stay Put." Here, he theorizes a right to stay in place despite the machinations of the housing market – and pushes for ridding the housing market of the profit motive all together, in order to make stable, decent housing accessible to all. He recommends policies to protect low-income home owners (including limits on property tax increases and challenges to shady mortgage lending practices) and renters (including enacting "just cause" eviction statutes and rent controls) (HARTMAN 1984). The right to stay put, here and in future works (see BRATT, STONE et al. 2006; HARTMAN 2006), is interchangeable with a right to housing. But a right to stay put should extend beyond the right to simple shelter. Being able to remain in place is about being able to continue to participate in the many daily rhythms of place, of which housing is just one part. A right to stay put, that is, can be theorized as part of a larger right to dwell.

The difference between dwelling in a place and simply being housed there is, in part, that dwelling requires time and intimacy. Dwelling, according to philosopher Martin HEIDEGGER, means to remain in a place over time (HEIDEGGER 1971). Dwelling, the environmental psychologist Susan SAEGERT writes later, connotes a certain intimacy that simple housing does not necessarily convey (SAEGERT

1985). Indian architect and activist Jai SEN articulates the distinction in his essay "It's 'Dwelling,' Stupid, not 'Housing'!" He writes:

> '[H]ousing' – as it is commonly understood – refers merely to the four walls and roof (and floor) in which we may dwell, the building, whereas 'dwelling' refers to something far deeper: To the existential relations of *living in a place*, and to the wider social and cosmo-logical meaning of this action. It refers to the act of 'settling and residing' somewhere, of in-*habiting* it, and of making it one's home; and ultimately, of struggling for and building 'one's place in the world'. And – I propose – it is this (and not the gaining of the mere object called 'housing') that we, as living, sentient beings, all really struggle for (SEN 2002, 1–2, emphasis in original).

Basic housing is still a demand in much of the world, and much of the United States as well. But there are other necessary elements of daily life, as SEN suggests, that the bare concept of housing does not include. In the Maple Plains[1] neighborhood of Washington, D.C., where gentrification has been studied several times since the mid-1970s, there is continual concern about the displacement of people from housing (GALE 1976; WILLIAMS 1988; MODAN 2007). But there is also a concern for the displacement of culture and music from the streets, and of conversation and communication to the internet (LOUGHRAN 2008). In 1998, a small group of Maple Plains residents decided to fight against the displacement of their neighbors' voices and cultures from a community dialogue that was in-creasingly shaped by internet discussion boards, and dominated by a few privi-leged interests. They wanted to create new ways to communicate and build understanding among residents, which they hoped would lead to a more just neighborhood: a more inclusive place in which to dwell. To achieve all this, they decided to start a microradio station.

THE PARTICULAR GEOGRAPHY OF MICRORADIO

A typical full-power radio station blankets a city with its sound. Regardless of where you may be in a given metropolitan area, you will hear its signal, and you can assume it will always be there. A microradio station is different: its signal is much less strong, and its broadcast reach is much smaller. "Microradio" is not a precise term, but rather a loose description for a radio station that is, simply, very small. Such a station may cover a very small town, or one neighborhood, or even just a single apartment building. The typical full-power FM station might broad-cast at 50,000 watts, while a microradio FM station may broadcast anywhere from one tenth of a watt to a hundred watts.

Microradio in the United States emerged in part as a response to a radio in-dustry that has, since its inception, moved ever more towards commercialization and homogenization. In the first few years of the 20th century, radio stations flourished as experiments, and at first no one could figure out how to make a

---

1   The name of the neighborhood has been changed.

profit from broadcasting (MCCHESNEY 1998). After World War I, the company RCA, which manufactured radio equipment, was the only entity making money off radio; it encouraged the wild proliferation of amateur stations, since station operators needed to buy RCA gear in order to broadcast (SOLEY 1999). But by the time of the creation of the Federal Communications Commission (FCC) with the Communications Act of 1934, commercial interests had realized that broadcast radio in fact offered great opportunity for profit. They convinced the government that radio supported by advertising dollars would ensure the greatest freedom of expression, and the commercial CBS and NBC networks began to dominate the airwaves (MCCHESNEY 1998). Still, the 1934 Act limited concentration of station ownership, banning any single individual or company from owning more than two stations, one AM and one FM, within a single market (GREVE, POZNER et al. 2006).

Starting in 1948, the FCC began issuing Class D licenses for FM radio, which at the time was unprofitable compared to AM radio. A Class D license allowed noncommercial operation at 10 watts for universities, and later, for other noncommercial entities. The small wattage of Class D licenses meant that they could be founded and run for much less money than full-power stations, and were therefore more accessible to small groups with small budgets (SOLEY 1999). In 1967, the Public Broadcasting Act created the Public Broadcasting System (PBS) and National Public Radio (NPR), scoring a win for noncommercial media (MCCHESNEY 1998). But in 1978 the FCC stopped issuing Class D licenses, after several years of lobbying from NPR and the National Federation of Community Broadcasters, groups that represented larger-scale noncommercial broadcasters, who complained that the tiny 10-watt stations cluttered up the airwaves and were an inefficient use of the small part of the FM spectrum that is allocated for noncommercial use (WALKER 2001). In addition, existing Class D stations were given a secondary status that meant they could be displaced by larger stations (OPEL 2004). Class D stations continued to operate, but slowly, their numbers sank. Meanwhile, ownership restrictions continued to loosen under lobbying pressure from the commercially-oriented National Association of Broadcasters (NAB), which expressed concern about the dwindling profitability of radio. In 1985, restrictions were relaxed such that entities could own up to twelve AM and twelve FM stations nationwide (GREVE, POZNER et al. 2006).

But it was the Telecommunications Act of 1996 that really removed restrictions on media ownership, and made owning a radio station potentially much more profitable. Before the law was passed, an individual or company could own up to two FM and two AM stations within a single market area; afterwards, an entity could own up to eight stations within a single market, and could own an unlimited number of stations across the country. Within the first year of the law's passage, twenty percent of all radio stations changed ownership in a "wave of consolidation" (WIKLE and COMER 2009, 369). By 1999, nearly half the nation's 11,000 stations had been sold, almost always from smaller firms to larger firms (RUGGIERO 1999). By 2003, a single company, Clear Channel Communications, owned enough radio stations to reach 70% of American listeners (GREVE, POZNER et al. 2006). Investors were developing nationwide chains of stations, cutting costs by mass-manufacturing programming for dozens of stations in a single studio in an anonymous city (BALLINGER 1998; WALKER 2001).

It was during this frenzy of radio ownership consolidation that microradio in the United States began to flourish in direct opposition to the increasing commercialization and homogenization of radio. Activists realized they could build their own small stations from a few hundred dollars' worth of electronics, search the FM dial for an empty frequency, and take to the air. Because no low-wattage FM radio licenses were available from the FCC, it was impossible to apply for one, so microradio was a strictly an illegal, "pirate," affair. Scores of stations multiplied across the country: Radio Mutiny in Philadelphia; KIND Radio, in San Marcos, Texas; Steal This Radio, a tiny station on New York's Lower East Side; Iowa City Free Radio; Beat Radio, a dance station in Minneapolis; Grid Radio in Cleveland; Free Radio Berkeley, which helped many other stations get started; and many, many more. Microradio was as wild hodge-podge of right wing gun enthusiasts, left wing social justice activists, zealous small businessmen, Christian evangelists, and anarchists. Some stations mimicked mainstream commercial radio, and some broadcast hell and damnation straight through the day and night. But the most interesting were those that crafted programming that reflected their communities' cultures and languages, and opened up room for anyone to come on the air and discuss whatever they wished. Free Radio Memphis had a show called "Solidarity Forever," focused on local labor issues and hosted by a member of the Industrial Workers of the World. Radio Mutiny had a show hosted by the Condom Lady, a public health worker by day who dispensed safe sex advice over the airwaves by night. Steal This Radio had an open mic hour in which anyone could come in off the streets and join in the broadcast. Beat Radio was started by a former commercial DJ in order to play dance music that other stations refused to play. Radio Zapata in Salinas, California, catered to the migrant workers of that town, ceasing operation when they left town for the winter and firing back up when they returned to work the fields in the spring (SAKOLSKY and DUNIFER 1998; SOLEY 1999; WALKER 2001). Though the FCC cracked down, shutting down more than 250 unlicensed stations in 1998 alone, an estimated 1,000 other stations continued to operate (OPEL 2004).

A station that operated out of a housing project in Springfield, Illinois, is widely considered to be the initial inspiration for the U.S. microradio movement, and demonstrates the unique geographic specificity of microradio (LANDAY 1998; SOLEY 1999). The John Hay Homes, built in 1940, was one of the first public housing projects in Illinois (DAVIS 1997). In the mid-1980s the project, which was almost entirely African American, was home to about 3000 people – one fifth of the black population of Springfield (SOLEY 1999). In 1986, the project's Tenants' Rights Association decided they needed to start their own radio station in order to reach residents. Mbanna Kantako, a blind man who had grown up in the project and lived there with his family, took the lead on developing the station, which was initially named WTRA to denote its connection with the tenant association.

WTRA started off with a single watt of power – tiny, but enough to cover the entire housing project and reach much of the black population of highly segregated Springfield. The tenants association used a small grant to buy the equipment and didn't worry about being busted by the FCC. Kantako ran the station out of his apartment with his wife, Dia, and their children. They played music, made political commentary, and, aware that much of their audience was functionally illiterate, read books and newspapers aloud over the air. They interviewed victims of police brutality in and around the project, and rebroadcast police dispatches so residents would know where the police were and what they were doing at any given time (SOLEY 1999). Kantako, who also helped organize the tenant association's 1000-volume Malcolm X Children's Library, described his station as "'a Black Panther political education class on the radio'" (quoted in LANDAY 1998, 94). Over the years, the station's name changed, from WTRA to Zoom Black Magic Liberation Radio to Black Liberation Radio to African Liberation Radio to Human Rights Radio. Of the name changes, Kantako explained:

> We're learning as we go…We named our original organization the name we thought was a solution to our problems – the Tenants Rights Association. We thought if we got tenants' rights – boom – everything would fall into place. We learned that wasn't the case' (quoted in SOLEY 1999, 74).

In 1995, the Springfield Housing Authority decided to close the John Hay Homes in order to demolish it under the federal HOPE VI program (DAVIS 1997). Kantako responded by launching a new radio program called "The Great Land Grab," in which he argued against the destruction of the housing project and the "Negro removal" policies of the housing authority. But by mid-1996, the project was mostly vacant; in 1997, the Kantako family was finally moved out, and the John Hay Homes was destroyed. Though Kantako and his family continued to broadcast from their new home, their listeners had been scattered, and a single microradio station could no longer reach them all (SOLEY 1999). The power of the small station had been bound up in the physical concentration of its listeners, an intimate geography destroyed by policies that some activists decried as intentional "spatial deconcentration" strategies designed to dilute black political power (see Yulanda Ward Memorial Fund 1981). But even though the station's original home was destroyed, a small network of Black Liberation Radio had already begun to flourish, inspired by Kantako's example, with stations taking to the air in Decatur, Illinois; Chattanooga, Tennessee; and Richmond, Virginia (SOLEY 1999).

Japan's "mini FM" movement was another manifestation of hyperlocal radio. The movement began in the early 1980s as an explosion of tiny unlicensed radio stations. Tetsuo KOGAWA, an activist, professor, and leader in the mini FM movement, was interested in the communicative and performance art possibilities of small-scale radio, and had been inspired by Italy's free radio movement. One of the earliest mini FM stations was Radio Home Run, founded by KOGAWA's former students in a bohemian district of Tokyo. According to KOGAWA, the station was named for "'a baseball term but its connotation was to 'cross distant borders,' because they wished to cross the borders of every obstacle (not only the airwaves regulations but also sociocultural differences)'" (CHANDLER and NEUMARK 2005, 199). KOGAWA believed that the seeming limitation of the mini FM stations' tiny wattage allowed them to serve a unique community function. These stations, he theorized, should be intentionally miniature, with a broadcast range small enough that anyone listening could bicycle over within a few minutes to join the conversation (SOLEY 1999). "Paradoxically," KOGAWA writes,

> limitations can always transform negative elements into positive ones. In our experience, listeners frequently visit their neighborhood stations, which consequently become communal gathering places. Given its essential difference from mass media, this should be the most positive function of free radio (KOGAWA 1993, 94).

For KOGAWA, the beauty of mini FM was that it erased the border between the producer and the listener of radio. And while the broadcasts that could be heard over the air were important, what were just as important were the in-person interactions that the small geographic scale of the broadcasts encouraged.

Locally-oriented radio was what the Swedish geographer Torsten HÄGERSTRAND had in mind when he theorized the "possibility space" of a communication technology. Radio, he argued, had been used to centralize the production and distribution of information and culture, but there was no reason the medium could not be used to enhance local, "situational" knowledge instead. He writes:

> [T]elemedia have an inherent tendency to promote hierarchical and centre-directed links resulting in the withdrawal of people from face-to-face communication. But these limitations do not totally circumscribe the 'possibility space'…To bring broadcasting down from the national to the regional level implies that nearness might after all be of some importance. The nature and sources of situational knowledge form the interesting side of the matter. General knowledge is the same everywhere, but situational knowledge is bound to place and time (HÄGERSTRAND 1986, 20).

RADIO CPR

Radio CPR (short for "Community Powered Radio") was founded in Maple Plains in 1998 to create the kind of possibility space HÄGERSTRAND theorizes. It reflects elements of Human Rights Radio, Radio Home Run, and the many microradio stations that have flourished across the country in recent years. It has a small broadcast radius, covering only a few densely populated neighborhoods, and its members generally recognize that its strength is in its tight geographic focus. As one of the station's founding DJs, Aphrodite, told an interviewer in 2005, "'Hopefully we have a radio station where anyone [elsewhere] in the world would have no interest in what we're talking about'" (quoted in BRINSON 2006, 552).

The germ of the idea for the station sprouted in 1996 when two Maple Plains social workers, outraged by that year's federal welfare and immigration "reform" laws, founded a group called Stand for Our Neighbors. They were witnessing the new legislation's immediate negative impacts on their clients, and wanted to organize to build solidarity with poor people in the neighborhood. Maple Plains was one of the most diverse parts of the city, with people of different classes, races and languages living within close proximity and passing each other on the street constantly. But people did not necessarily talk to each other in any substantive way, or have the opportunity to get to know one another. Increasingly, neighborhood conversation took place via internet discussion boards, which excluded swaths of the population and tended to degenerate into less-than-compassionate rhetoric (MODAN 2007). In response, Stand for Our Neighbors organized events across cultures to encourage Maple Plains residents to get to know one another. These events ranged from "neighborhood cabarets" at local restaurants and churches, highlighting the diverse talents of local people, to public forums in the basement of the neighborhood public library.

At one forum on crime and safety in the library's basement, teenagers studying radio production at a local youth center played tapes of interviews they had conducted with other young people, on what safety meant to them. Their recordings were revealing. The young people, all Latino and African-American, had very different safety concerns than the mostly white, mostly upper-middle class people attending the forum. They did not express fear of being mugged, or of their cars being stolen, or of their homes being broken into. They were afraid of being harassed by police when they walked across the road that divided their poor neighborhood from the wealthier one to the west. They were afraid that their families would not be able to pay the rent, and that they would be evicted. They were afraid their family members would be deported. Safety was knowing they could walk down the street without being treated like criminals, that their families would not lose their homes, and that they would not be forced to leave the country. Watching the people at the forum listening to the tapes, it became clear how powerful it would be if these teenagers' voices could be heard outside the context of a neighborhood meeting in a library. A radio station could broadcast the taped interviews and the neighborhood cabarets, bring in local activists for live discussion, play music created in the neighborhood but never heard on the city's other radio stations, and serve as an outlet for every conceivable form of auditory creation. And so Radio CPR began.

Since its first broadcast in October 1998, Radio CPR has hosted scores of shows and hundreds of DJs. The station broadcasts every night, starting at about 6:00 PM, and signing off at about 1:00 AM. A typical Wednesday night's broadcast in the summer of 2010, for example, begins with a show hosted by three librarians, in which they play music checked out from the neighborhood library, and promote public library events; it's followed by a longrunning show hosted by Paleface, a British expatriate with a wealth of old reggae and punk records, who tells terrific tales of seeing punk bands in London in the '70s; and the evening finishes up with Zombie, another longstanding DJ whose show is devoted to go-go, one of the city's uniquely indigenous musical forms that is wildly popular yet barely represented on the city's airwaves. Over the years, CPR programming has included a Saturday morning show in which a neighborhood dad read children's stories and played kids' music; the Sunday evening "Neighborhood Power Hour," which for ten years relayed neighborhood news, interviewed local activists, and was occasionally turned over to one of the DJ's teenage sons and his friends to play punk records (the DJ mom often referred to the station as a "gang prevention program" because of its ability to keep her son and his friends out of trouble); a Thursday evening experimental music show in which local musicians brought their instruments into the station and improvised live on the air; a Friday evening youth show, in which young people interviewed each other about the public school system, teen sex, and other issues, and read their poetry and played their favorite songs; a show devoted exclusively to Screw, the weird form of sloweddown rap originating out of Houston in the '90s; "The People's Music Hour," a folk music show interspersed with politics, hosted by a longtime neighborhood couple; numerous bilingual Spanish/English talk shows; and much, much more. Radio CPR currently broadcasts at about seventy watts, which covers two or three neighborhoods. It has never received any formal funding, and operates on the barest of shoestring budgets, supporting itself exclusively through small benefit concerts, dance parties, and record sales. Though between forty and fifty people participate regularly as DJs, a smaller group tends to be at the core of decision-making. Members gather at monthly meetings and do work in smaller committees that focus on technical and programming issues. Keeping the station going requires a lot of work, but the ability to create material sound where none existed before is powerful motivation.

From the outset, it was clear that Radio CPR was going to have to walk a fine line between being inclusive and expansive. Because of the station's undocumented nature, its members needed to be careful about whom to involve, while not succumbing to paranoia about being busted by the FCC. The station was founded by a group of friends who knew that they both needed to involve people they trusted and also involve people they that did not yet know. The solution was to make new friends and expand friendship circles outward as more and more trust was built with new people, and connections were strengthened. As of this writing, twelve years after its founding, the station has never been contacted by the FCC, and working together on the station has engendered many friendships across race, class, religion, immigration status, and sexuality.

The station's friendship circles have expanded out in all sorts of surprising ways. One evening, for instance, I walked into the neighborhood 7-Eleven to buy a bottle of orange juice. I was carrying a portable tape recorder and microphone, and was on my way to do an interview for my show. The 7-Eleven in our neighborhood is staffed almost entirely by Somali immigrants – perhaps the better term is war refugees (Maple Plains has long provided a home for people displaced from war and its repercussions throughout the world, from El Salvador to Vietnam). A long scar ran across the neck of the Somali man working the counter, slicing from jaw to collarbone, and a piece of his left ear was missing. He grinned as he rung me up, and asked what I was doing with the tape recorder. I told him about the station, and encouraged him to join us. He was interested, and we gave him a show in which he and a friend played Somali music and debated Somali politics – all in Somali, so I never understood a word. But the workers started listening to Radio CPR in the 7-Eleven, and walking into that chain store late at night and hearing our signal wafting through its fluorescent aisles was thrilling.

FIGHTING DISPLACEMENT WITH MICRORADIO

A radio station is not a house. It does not provide material shelter. But there are particular ways in which microradio's unique geography can aid communities in fighting displacement.

First, a microradio station can air programming that directly deals with the problems of housing displacement. It can host shows in which DJs discuss tenants' rights and housing concerns, interview tenant leaders, lawyers and activists, and share news of housing organizing efforts in their neighborhood and across the city. In 2003, for example, the tenants of a Maple Plains apartment building were able to buy their building from their landlord, a notorious slumlord, and preserve it as affordable housing in perpetuity. The tenants' association, which was highly organized and translated every document and every meeting into Spanish and Vietnamese in order to ensure communication among all tenants, was presided over by a Haitian immigrant who had moved into the building in 1978 (MORENO 2003). She came to the Radio CPR studio on multiple occasions to discuss the building's case on the air, and to ask for neighborhood support. In 2005 the tenants of another neighborhood apartment building were fighting condominium conversion, which would have displaced many of them (COHN 2005); again, the tenants' association president came on Radio CPR to discuss the case, and ask listeners to testify at the city council in their support. In 2008 an apartment building on Maple Plains' main commercial strip burnt to the ground in a terrific blaze, after having racked up over 7000 housing code violations in recent years; 200 low-income, mostly immigrant residents were instantly displaced (DVORAK and KLEIN 2008). Radio CPR DJs were quick to report on the disaster and offer assistance to the suddenly homeless families. The highly localized nature of the radio station ensured that listeners would be interested in the fate of the buildings they passed by every day, and also made it easier for tenant leaders to walk the few blocks over to the studio to share their stories.

Second, a microradio station can air music and culture that has been displaced from other media outlets. Go-go, punk, and bluegrass are all musical forms that either originated in D.C. or have flourished there in unique ways, but over the years it has become harder and harder to find this music on the radio dial. In 2001, for instance, WAMU, one of the city's two public radio stations, decided to cut its afternoon bluegrass and classic country. Washington, once known as the "Nashville of bluegrass," had long been an important center for bluegrass musicians, and the radio had been one way for musicians and fans to remember their history and keep up with current musical trends. But the WAMU manager said her listeners were more interested in news than music, so she replaced the bluegrass and country programming with more news and talk, essentially mimicking what the city's other public radio station, WETA, was already airing at that time, thus further eroding the diversity of the city's airwaves (AHRENS 2001). Programming at larger radio stations, even noncommercial ones, tends to follow financial dictates, and music that can't sell ads or generate listener donations may not make the cut. But the scale of microradio is so small, and its financial needs are consequently so minor, that it allows all sorts of musical forms to flourish. One of Radio CPR's earliest DJs was herself a player of old time, blues, and bluegrass, who led the music at her Maple Plains church and in the 1970s had produced a beautiful album of songs, featuring a cover photograph of her standing on Maple Plains' main street. Her husband had written the record's liner notes. Together they aired music of the people, including the songs of many folk and bluegrass musicians. Go-go and punk also proliferate on the station, and local musicians consistently receive airplay, at times performing live on air.

A microradio station can also air music and culture that has been displaced from the neighborhood's streets themselves, and work to bring that music and culture back. In the late '90s, a small group of Maple Plains neighbors pressured local businesses to stop hosting live music, DJ-ing and dancing. They professed fear that the neighborhood was turning into an entertainment district, and felt threatened by the possibility of a noisier commercial strip. They warned that if the businesses refused to stop hosting music, their group would contest their liquor licenses (this was not an empty warning: the head of the neighbors group was a former member of the city's Alcoholic Beverage Control Board). The neighborhood businesses, which over the years had hosted live country, rock, punk, and Latin music, were upset about the pressure, but fearing the loss of their liquor licenses, they complied. A blow felt particularly hard by the neighborhood was the resultant loss of mariachi bands, which had once roamed from restaurant to restaurant, sometimes playing their instruments as they walked down the street, or stopping to play a song in the small neighborhood plaza (FISHER 2007). Members of Radio CPR responded to this de facto ban on live music by working for several years with other neighborhood residents to bring musical performance back to the neighborhood. (Maureen LOUGHRAN's excellent (2008) dissertation, *Community Powered Resistance: Radio, Music Scenes and Musical Activism in Washington, D.C.*, focuses on Radio CPR's work on this issue.) Because of this work, bands are once again permitted to play, DJs are permitted to spin records, and people are permitted to dance in Maple Plains. Radio CPR DJs now host regular cumbia/tropicalia music dance nights at a local Salvadoran restaurant; one of their recent dance nights was a benefit to send several CPR DJs to the Allied Media Conference, an annual convergence of progressive media-makers.

The tropicalia dance nights point to another way that microradio can help in the fight against all types of displacement: by creating physical spaces where people can both work and party together. Most radio listeners listen to the radio alone, in their cars or in their homes. Listening to the radio alone can be an intimate experience: the best DJs sound like they are speaking to you and you alone. But microradio is not just about what is heard beaming out over the airwaves, but about the work that goes into creating that signal. It is necessarily a communal project that relies on the freely given time and energy of its participants. Experience with Radio CPR shows that when people work together on a common project in which they share a common passion, they get to know each other in deep ways. They learn how to communicate, and they learn to trust each other. And they build social networks that extend far beyond their original groups of friends. These networks are built among DJs during monthly meetings at local bars, afternoons spent cleaning the studio, rebuilding the transmitter or repositioning the antenna, and evenings on front porches combing through applications for new shows. Networks are built among DJs and other members of the community during the Maple Plains annual street festival, where Radio CPR organizes the children's stage every year; the annual neighborhood Halloween block party, where CPR provides live spooky music over the radio, and DJs serve as judges for the costume contest; local neighborhood council meetings, where DJs speak up in support of policies that support small businesses and immigrants' and housing rights; and dance parties and benefit concerts, where local bands play and people socialize into the night. It is these networks that help build a community that is stronger in the face of displacement, be it of home, music or public culture.

Finally, microradio may aid the fight against displacement most fundamentally in the way it challenges the private enclosure of public resources. The Radio Act of 1927 stipulated that broadcasters operate in the public interest, that licenses be renewed every three years, and that the airwaves were public property, owned by the U.S. government (SOLEY 1999). But from the inception of radio's regulation, elected officials have given away the spectrum to private industry: both in the hope that large private companies could make most efficient use of it, and because, since the early 1930s, defying the interests of big media has proven to be political suicide (MCCHESNEY 1998). Similarly, the history of land ownership in the U.S. has been one of expropriation, most fundamentally the original expropriation of land from its Native residents. To engage in microradio is to reclaim airspace that was once considered to be part of the common realm. To fight to stay put is to defy a system of property ownership that prioritizes protecting profit over meeting human need. Such struggles can take place within legal frameworks: residents can push governments to pass laws that protect and create affordable housing, just as media activists, Radio CPR members among them, successfully pushed the FCC in 2000 to create a new noncommercial low power FM (LPFM) community radio service (AHRENS 2000; OPEL 2004). But these struggles can also take place outside the law. HARTMAN ends his 1984 essay by commending squatters who take over buildings in order to house the homeless. An analysis of the community stations licensed under the initial rounds of the LPFM service found that, despite tremendous efforts to help progressive groups get licenses, most new stations went to rural, white America, and a large number of licenses went to Christian broadcasting organizations (WIKLE and COMER 2009). Legislation moving through Congress as of this writing would expand the service to provide opportunities for more low power stations, and in more urban areas, but its passage is not guaranteed. The opportunities presented by a legal low power FM service are great, and represent a major victory by media democracy groups. But squatting the airwaves is still important: it makes a political point about the continued need to fight a telecommunications system that is based on corporate profit, while at the same time directly creating a means of communication.

Fighting displacement with microradio is in no way unique to Radio CPR. Steal This Radio, for example, was simultaneously squatting the airwaves and the land, operating out of a squatted building on New York City's Lower East Side. That station covered the evictions of fellow squatters, fought against the "rent-slavery economy," and provided a space for countering the "numbing effect of the corporate mediascape we all inhabit [that] casts a long shadow over so much of the lived experience of the city" (DJ TASHTEGO 1998, 136).

THE FETISH OF PLACE, AND OF THE RADIO FORM

It's exciting to take the air into your own hands and make something new of it. DJ TASHTEGO calls it the "sweet mystery of radio," and it is a remarkable and magical thing (DJ TASHTEGO 1998, 133). But microradio is not immune to critique.

One critique is that it encourages a fetishization of place: a sort of hyper-local navel-gazing and obsession with one particular place. In its intense focus on a single small place, microradio may contribute to the same fetishization of neighborhood that plays such an important role in gentrification. Since Radio CPR's first broadcast, median home sales prices in Maple Plains have more than doubled in constant 2009 dollars (from $297,000 in 1998 to $621,000 in 2009), and the median income of mortgage borrowers has risen by 42% (Urban Institute and Washington DC LISC 2009). When the neighborhood gentrifies, access to the station's airwaves becomes a point of privilege. Not everyone can afford to live within the broadcast range. Being able to tune into a cool little neighborhood radio station might be just one more thing that makes living in Maple Plains special and desirable, and ever more commodified – possibly, in a contorted way, even contributing to displacement.

Another critique is that, increasingly, microradio relies on a fetishization of the radio medium itself. FM radio may be of waning consequence in a country in which most people have ready access to fast internet connections. One recent listing of online radio stations provides links to 14,000 of them, and this list is, apparently, far from complete (TAUB 2009). FM radio may be headed the way of LP records: a few people might obsess over the medium and think it's cool, but no one really needs it. The focus on one small place that microradio requires may be becoming just a quaint, nostalgic throwback.

These critiques merit thoughtful response. While Radio CPR certainly focuses mostly on the neighborhoods within its small broadcast range, it also encourages a proliferation of community media engagement throughout the city (and, through participation in events like the Allied Media Conference, throughout the country). Rather than expand their own station's wattage, CPR members have worked to connect with other groups in the Washington area, from D.C.'s Anacostia neighborhood to the town of Frederick, Maryland, that might be interested in starting their own stations, in order to encourage a local network of microradio. This is a vision of microradio self-reproduction inspired by Free Radio Berkeley's Stephen DUNIFER's call to "let a thousand transmitters bloom" (SAKOLSKY and DUNIFER 1998). It is potentially a manifestation of what Ron SAKOLSKY calls "rhizomatic radio:" "[r]ampaging sound wave tubers where each stem is itself a rootstock emitting new roots everywhere along its sonic path" (SAKOLSKY 1998, 7). In terms of Radio CPR's relationship with its gentrifying neighborhood, the station is not necessarily fixed forever in space. Members may at some point decide to move the studio to another neighborhood in which more of its members live, or engage in more mobile broadcasts in other neighborhoods, in order to keep from being fossilized in place as a relic of a bygone era.

As for FM's continued viability as a medium, it is important to remember that a microradio station does not exist in a communicative vacuum. As Felix GUATTARI, who was involved in Italy's pirate radio station movement in the 1970s, writes of those stations:

> We realize here that radio constitutes but one element at the heart of an entire range of communication means, from daily, informal encounters in the Piazza Maggiore to the newspaper – via billboards, mural paintings, posters, leaflets, meetings, community activities, celebrations, etc." (GUATTARI 1993 [1978], 86).

The work of the neighborhood radio station is not just about the FM radio signal, but about the context in which it is created and heard. The handmade radio station – more mystical and seductive than the increasingly mundane internet – can be a hub around which all sorts of communication spins. There also may be something important about the nature of unlicensed radio itself: it may be because such stations are undocumented and therefore cannot receive any funding to operate that it can retain the kind of radical edge that community groups dependent on foundation grants – including groups working against displacement – may eventually lose (see INCITE! Women of Color Against Violence 2007). Along these lines, it is useful to consider future possibilities for microradio.

CONCLUSION: MICRORADIO AS CO-RESEARCH

Microradio is in a very different position today than it was in 1998, when Radio CPR was founded and hundreds of other tiny pirate stations were taking to the air across the country. The LPFM service has channeled some of the energy of the late '90s into licensed radio stations, and the terrific rise of the internet has totally altered the geography of media. The problem of displacement, however, has not changed, and social scientists still argue over its scale and impact (see FREEMAN and BRACONI 2004; and the response of NEWMAN and WYLY 2006).

I see the future of microradio as part of a larger project of neighborhood-based co-research. Co-research, also known as militant research or workers' inquiry, is research that seeks to break down barriers between researchers and the object of research, and between research and politics. An early example is Karl MARX's 1880 call for "A Workers' Inquiry," published in *La Revue Socialiste*, in which he called for French workers to systematically investigate their own conditions in order to, as he put it, "prepare the way for social regeneration" (MARX 1880, 1). In this piece, Marx introduces a one hundred question survey, which include questions ranging from "What is your trade?" to "Has the government or municipality applied the laws regulating child labor? Do employers submit to these laws?" to "Have you noticed, in your personal experience, a bigger rise in the price of immediate necessities, e.g. rent, food, etc., than in wages?" (MARX 1880, 2–5). These questions were designed to elicit facts as well as raise consciousness. The feminist consciousness raising groups of the 1960s and '70s have also been theorized as a form of co-research, in that participants investigated their own experience of the world in order to change it (MALO DE MOLINA 2004). As feminist Kathie SARACHILD wrote of consciousness-raising in 1973:

> The decision to emphasize our own feelings and experiences as women and to test all generalizations and reading we did by our own experience was actually the scientific method of research…It was also a method of radical organizing tested by other revolutions (quoted in MALO DE MOLINA 2004, 5).

Neighborhood-based research has proven to be a powerful component of anti-displacement work. HARTMAN et al. highlight, for example, the research-intensive work of San Francisco's Duboce Triangle Neighborhood Alliance. Alliance members studied census data, real estate listings, and city and county records for the Duboce Triangle neighborhood in order to gain a better understanding of the nature and extent of displacement, and followed up their statistical analysis with a door-to-door survey of neighborhood residents. They found that, between 1970 and 1978, housing prices had increased far beyond incomes, and that working class residents were moving out and young professionals were moving in (HARTMAN, KEATING et al. 1982). Similarly, the Washington Urban League's *SOS '78 – Speak Out for Survival* door-to-door survey conducted in 1978 in four D.C. neighborhoods, including Maple Plains, found that 43% of those who had moved in the past two years cited rent increases, evictions, or urban renewal as reasons for moving (GALE 1980). In both San Francisco and Washington, D.C., neighborhood-based research in the late '70s countered the prevailing federal narrative that displacement was not a problem, and provided a basis for demanding changes to city policies on land, housing and real estate.

A microradio co-research project could serve as a repository and a hub for research that aids anti-displacement struggles, and a way to communicate information discovered. Microradio already has a long history of co-research, manifested in part by the Black Liberation Radio network of stations. Members of these stations were intimately aware of the injustice of land and housing distribution, and addressed those injustices on the air. Because microradio stations of necessity are so tightly geographically focused, they are in a good position to serve as a center for neighborhood co-research. At Radio CPR, as at many other microradio stations, this research has already been happening in informal and ad hoc ways, through in-studio interviews, discussions, and reporting on local issues of housing and land. But this work could become more intentional. The challenge is to deepen and extend the "possibility space" of radio in order to continue to work for a neighborhood – and a world – that values the right to dwell, and the right to stay put.

REFERENCES

AHRENS, F. (2000): Political Static May Block Low-Power FM; FCC, Congress Battle Over Radio Plan. *The Washington Post*. Washington, D. C.

AHRENS, F. (2001): WAMU Cuts Daily Bluegrass, Country Shows. *The Washington Post*. Washington, D.C.

BALLINGER, L. (1998): Broadcast Confidential. SAKOLSKY, R. and S. DUNIFER (Eds.): *Seizing the Airwaves: A Free Radio Handbook*. San Francisco, AK Press, 25–28.

BRATT, R. G., M. E. STONE et al. (2006): Why a Right to Housing is Needed and Makes Sense: Editors' Introduction. BRATT, R. G., M. E. STONE and C. HARTMAN (Eds.): *A Right to Housing: Foundation for a New Social Agenda*. Philadelphia, Temple University Press, 1–19.

BRINSON, P. (2006): Liberation Frequency: The Free Radio Movement and Alternative Strategies of Media Relations. *The Sociological Quarterly* 47, 643–568.

CHANDLER, A. and N. NEUMARK (2005): Mini-FM: Performing Microscopic Distance (an email interview with Tetsuo Kogawa). CHANDLER, A. and N. NEUMARK (Eds.): *At a Distance: Precursors to Art and Activism on the Internet*. Cambridge, MIT Press, 190–209.

COHN, D. V. (2005): Fighting Sale, Renters Find New Allies at City Hall; District Gets Assertive Over Condo Conversion. *The Washington Post*. Washington, D. C.

DAVIS, J. (1997): Rebuilding. *Illinois Issues*. 23, 20–23.

DJ TASHTEGO (1998): Community Struggle and the Sweet Mystery of Radio. SAKOLSKY, R. and S. DUNIFER (Eds.): *Seizing the Airwaves: A Free Radio Handbook*. San Francisco, AK Press, 133–141.

DVORAK, P. and A. KLEIN (2008): D.C. Blaze Displaces Nearly 200. *The Washington Post*. Washington, D.C.

FEINBERG, L. (1975): Housing Costs Outrun Income of Blacks. *The Washington Post*.

FISHER, M. (2007): A Counteroffensive on Mt. Pleasant's 'Voluntary' Music Bans. *The Washington Post*. Washington, D.C.

FREEMAN, L. and F. BRACONI (2004): Gentrification and Displacement: New York City in the 1990s. *Journal of the American Planning Association* 70(1), 39–52.

FRIEDEN, B. J. and A. P. SOLOMON (1977): *The Nation's Housing: 1975 to 1985*. Cambridge, MA, Joint Center for Urban Studies of MIT and Harvard University.

GALE, D. E. (1976): The back-to-the-city movement – or is it? a survey of recent homebuyers in the Mount Pleasant neighborhood of Washington, D.C. Washington, D.C., Department of Urban and Regional Planning, George Washington University.

GALE, D. E. (1980): Neighborhood Resettlement: Washington, D.C. LASKA, S. B. and D. SPAIN (Eds.): *Back to the City: Issues in Neighborhood Revitalization*. New York, Pergamon Press, 95–115.

GOLDFIELD, D. R. (1980): Private Neighborhood Redevelopment and Displacement: The Case of Washington, D.C. *Urban Affairs Review* 15(4), 453–468.

GREVE, H. R., J.-E. POZNER et al. (2006): "Vox Populi: resource partitioning, organizational proliferation, and the cultural impact of the insurgent microradio movement." *American Journal of Sociology* 112(3), 802–37.

GUATTARI, F. (1993 [1978]): Popular Free Radio. STRAUSS, N. and D. MANDL (Eds.): *Radiotext(e)*. New York, Semiotext(e), 85–89.

HÄGERSTRAND, T. (1986): Decentralization and Radio Broadcasting: On the 'Possibility Space of a Communication Technology.' *European Journal of Communication* 1(7), 7–26.

HARTMAN, C. (1984): The Right to Stay Put. GEISLER, C. C. and F. J. POPPER (Eds.): *Land Reform, American Style*. Totowa, NJ, Rowman and Allenheld, 302–18.

HARTMAN, C. (2006): The Case for a Right to Housing. BRATT, R. G., M. E. STONE and C. HARTMAN (Eds.): *A Right to Housing: Foundation for a New Social Agenda*. Philadelphia, Temple University Press, 177–192.

HARTMAN, C., D. KEATING et al. (1982): *Displacement: How to fight it*. Berkeley, National Housing Law Project.

HEIDEGGER, M. (1971): *Poetry, Language, Thought*. New York, Colophon Books.

HUD (1979): *Displacement Report*. Washington, D. C., U.S. Department of Housing and Urban Development.

INCITE! Women of Color Against Violence (2007): *The Revolution will not be Funded: Beyond the Non-Profit Industrial Complex*. Cambridge, MA, South End Press.

KOGAWA, T. (1993): Free Radio in Japan: The Mini FM Boom. STRAUSS, N. and D. MANDL (Eds.): *Radiotext(e)*. New York, Semiotext(e), 90–96.

LANDAY, J. (1998): "We're Part of the Restoration Process of Our People:" An Interview with Mbanna Kantako (Human Rights Radio). SAKOLSKY, R. and S. DUNIFER (Eds.): *Seizing the Airwaves: A Free Radio Handbook*. San Francisco, 94–99.

LEGATES, R. T. and C. HARTMAN (1981): Displacement. *Clearinghouse Review* 15(3), 207–249.

LOUGHRAN, M. E. (2008): Community Powered Resistance: Radio, Music Scenes and Musical Activism in Washington, D.C. *Ethnomusicology*. Providence, RI, Brown University. PhD.

MALO DE MOLINA, M. (2004): Common Notions, Part 1: Workers-inquiry, Co-research, Consciousness-raising. *Notas Rojas*.

MARX, K. (1880): A Workers' Inquiry. *La Revue Socialiste*.

MCCHESNEY, R. (1998): The Political Economy of Radio. SAKOLSKY, R. and S. DUNIFER (Eds.): *Seizing the Airwaves: A Free Radio Handbook*. San Francisco, AK Press, 17–24.

MODAN, G. G. (2007): *Turf Wars: Discourse, Diversity, and the Politics of Place*. Malden, MA.

MORENO, S. (2003): Tenants' Dreams Rescue NW Building; Low-Income Housing Option Preserved with $4 Million Renovation of Decrepit Site. *The Washington Post*. Washington, D. C.

NEWMAN, K. and E. WYLY (2006): The Right to Stay Put, Revisited: Gentrification and Resistance to Displacement in New York City. *Urban Studies* 43(1), 23–57.

OPEL, A. (2004): *Microradio and the FCC: Media Activism and the Struggle Over Broadcast Policy*. Westport, CT.

REINHOLD, R. (1977): Middle-class return displaces some urban poor. *The New York Times*. New York, 1.

RUGGIERO, G. (1999): *Microradio and Democracy: (Low) Power to the People*. New York, Seven Stories Press.

SAEGERT, S. (1985): The Role of Housing in the Experience of Dwelling. ALTMAN, I. and C. M. WERNER (Eds.): *Home Environments*. New York, 287–309.

SAKOLSKY, R. (1998): Rhizomatic Radio and the Great Stampede. SAKOLSKY, R: and S. DUNIFER (Eds.): *Seizing the Airwaves: A Free Radio Handbook*. San Francisco, 7–11.

SAKOLSKY, R. and S. DUNIFER (1998): *Seizing the Airwaves: A Free Radio Handbook*. San Francisco.

SEN, J. (2002): *"It's 'dwelling,' stupid, not 'housing'!"* New Delhi.

SOLEY, L. (1999): *Free Radio: Electronic Civil Disobedience*. Boulder, CO.

SUMKA, H. J. (1979): Neighborhood Revitalization and Displacement: A Review of the Evidence. *Journal of the American Planning Association* 45(4), 480–487.

TAUB, E. A. (2009): When the Web is in Reach, So is a Radio Universe. *The New York Times*. New York City.

THURSZ, D. (1966): *Where Are They Now? A Study of the Impact of Relocation on Former Residents of Southwest Washington who were Served in an HWC Demonstration Project*. Washington, D.C., Health and Welfare Council of the National Capital Area.

Urban Institute and Washington DC LISC. (2009): *Neighborhood Info DC*. Retrieved June 7, 2010, from http://www.neighborhoodinfodc.org.

WALKER, J. (2001): *Rebels on the Air: An Alternative History of Radio in America*. New York.

WIKLE, T. A. and J. C. COMER (2009): Barriers to Establishing Low-Power FM Radio in the United States. *The Professional Geographer* 61(3), 366–381.

WILLIAMS, B. (1988): *Upscaling Downtown: Stalled Gentrification in Washington, D.C.* Ithaca.

Yulanda Ward Memorial Fund (1981): Spatial Deconcentration. *Midnight Notes*, 28–31.
ZEITZ, E. (1979): *Private Urban Renewal: a different residential trend.* Lexington, MA.

Ryan J. Goode / Kate Swanson / Stuart C. Aitken

FROM GOD TO MEN: MEDIA AND THE TURBULENT FIGHT FOR RIO'S FAVELAS

> "…a single, irresponsible flow of images and feelings."
>
> Raymond WILLIAMS, 1974, 86

Directed by Fernando Meirelles and Katia Lund, *City of God* (2002) depicts the escalation of violent crime in a *favela* on the western outskirts of Rio de Janeiro from the late 1960s to the beginning of the 1980s. The movie focuses on the ways young people and, in particular, young children are enveloped in gangster life-styles. It looks at some youngsters' attempts to break away from those lifestyles and at police attempts to break the gangs. The main protagonist of the movie escapes the ravages of gang violence by becoming a photographer for a local newspaper. His ability to get candid photographs of notorious gang-members (his neighborhood pals) ingratiates the young photographer with the newspaper editors. The connection between media and the disturbances and transformations in Rio's *favelas* provides one point of departure for this essay. The second point of departure comes from the theme of *violence* and *staying put* as it is immortalized in the *City of God* tag-line: "Fight and you'll never survive…Run and you'll never escape." Our focus is the last two decade's fight for Rio's *favelas* and media culpability – for both good and bad – in that fight.

The spectacular success of *City of God*, and the subsequent television series and sequel film *City of Men*, is important for Rio in three ways. First, it coincides with the creation of the *Unidades de Policia Pacificadora* or Pacification Police Units (UPP) that began occupying the city's *favelas* from about 2008 and, second, the depiction of cleaned-up and gentrified *favelas* coincides with Rio's successful bid for the 2016 Olympic Games in 2009. The third reason that the success of *City of God* is important revolves around the part it plays in stimulating grassroots activism, including media activism, in response to the excesses of the UPP.

FIGHTING FOR THE RIGHT TO THE CITY: STREETS AND MEDIA

Welsh novelist and Marxist social critic, Raymond WILLIAMS (1974) spoke to the *flow* of media and the ways television's continuous sequencing affected him after a transatlantic trip to Florida. James HAYS (2009) points out that WILLIAMS' anecdotal quip about 'flow' has not been interrogated fully in deployments of the term by other media theorists and critics. Many dwell on the formal and aesthetic implications of flow while missing some of its affective and political consequences. Although WILLIAMS did not elaborate fully on feelings and politics in the book where the anecdote appears, he nonetheless notes that television is not just about technological change in communications and media, it is about change in socio-spatial assemblages that situate domesticity (privacy) increasingly at a distance from everything else while nonetheless connecting private sites with each other globally. He coined the term "mobile privatization" (1974, 20) to articulate the changing and geographically uneven interdependencies between social space and media space. The link between the "irresponsible flows" mentioned in our epigram and a regime of mobility and privacy suggests an important connection to the construction of networks of power that disenfranchise the urban poor. It holds up the image of gentrification in the global south wherein walled and gated communities provide secure privatopias for the urban rich who remain connected to the metropole, while the poor occupy increasingly marginalized and dilapidated housing, and unhealthy living conditions (LOW 2000; CALDEIRA 2001). It also suggests a control of knowledge through media that protects the rich and the private sphere. How these media are mobilized and used to further disenfranchise the poor is an important subject of this essay.

This essay is also about hope, and the provocation to act. Leftist Harvard law professor Mark TUSHNET (1984, 1363) critiques liberal theories of rights as a stultified moment that "capitalism's culture has given us." Alternatively, he argues that engagement should not be about rights but about how to fight for life, health and decent housing. In a provocative series of essays, Noel CASTREE and his colleagues (2010) note that democracy is something that only a small minority of the world's people enjoy, and the point is that this social and geographic inequity must change. The hopeful part of our essay comes from the ways grassroots activism enables disruption of corporate media. To the extent that corporate media is an irresponsible flow of images that generates certain kinds of feelings, we argue that a grassroot grasp on media, however tentative, creates turbulence in the flow. In the 1930s, Walter BENJAMIN (2008, 391) wrote about the problems of consumer society and the creation of dull, inarticulate masses. This "barbarism" must be reversed, he says, and in a later essay he talks about the ability of new media (for him, radio) to "not only orient knowledge toward the public sphere, but also simultaneously orient the public sphere towards knowledge" (BENJAMIN 2008, 404).

BENJAMIN's ideas are theorized more fully by Gilles DELEUZE (1986, ix) notion of the *affection-image,* which articulates an ontological connection between the virtual, the intensive and the actual. This is picked up by Arturo ESCOBAR (2008) when he argues for *redes* as networks or assemblages that open up the possible of transformative action in the face of blistering and relentless attacks by corporate and colonial capitalism. Life and social movements, he points out, are ineluctably produced in and though relations in a dynamic fashion…Images of *redes* circulated widely…in the 1990s [in the global south]…represented graphically as drawings of a variety of traditional fishing nets, lacking strict pattern regularity, shaped by use and user, and always being repaired, *redes* referred to a host of entities, including among others social movement organizations, local radio networks, women's associations, and action plans (ESCOBAR 2008, 26).

With this essay we focus on the *redes* of *Visão Favela,* a grassroots organization in the *favela* of Santa Marta (a *favela* that was particularly isolated thorough 'pacification'), and its attempt to wrestle control of popular media from the authorities who mobilize the UPP. BENJAMIN's notion that *truly popular interest is always active* comes together with DELEUZE's ontology and ESCOBAR's activism to bring clarity to how media work in Rio's *favelas*.

Santa Marta may be understood as the focus of Rio's fight for the right to the city. Henri LEFEBVRE wrote a convoluted and dense essay on the people's right to the city, which later became part of his *Writing on Cities* (1996). Arguing that the laboring body is central to the way cities work, LEFEBVRE noted that bodies have the capacity to create cities in which a wide range of desires, including sexual desires, are realized. LEFEBVRE's humanist hope was that people create cities that fulfill their bodily needs and desires, but his Marxist fear was that abstract representations deployed by architects' and planners' over-coded spaces create a de-corporealized city (HUBBARD 2006, 103). The context of gentrification and privatization further mummifies the corporeality of the people's city. What we will show below, is that the way Rio's *favelados – favela* residents – are pacified is very much about removing desire through sanitized control of music, radio and other media. Ironically, however, it is this very embodied passion and desire that Rio has capitalized upon in their 2016 Olympic media campaign. With the slogan, "Live your passion," the campaign celebrates the exuberance of global youth while squelching that same youthful exuberance amongst the city's urban poor. In a very DELEUZIAN and BENJAMINIAN sense, Santa Marta counters this ironic and authoritarian irresponsibility by creating desire and a corporeal city through counter flows of images, sounds and feelings that cause turbulence.

In what follows, we discuss the changes to Rio's favelas from the time of *City of God* and the creation of the UPP. The metaphor of Guiliani's sanitized *Times Square* becomes the paragon of a single, irresponsible flow of images and feelings that disrupts the lives of poor *favelados*. We look at the corporate media blitzkrieg around Rio's bid for the 2016 Olympics. We then turn to the ways authorities crack down on *favelados'* culture and media, with a particular focus on Santa Marta, and how *Visão Favela* resists this onslaught. We finish by bringing this work on media together with ESCOBAR's ideas about "epistemic borders," where activists produce knowledge and shuttle it "...back and forth alongside the modernity/coloniality, universality/pluriversality interface" (2008, 12–13). By so doing, we argue for a flexible account of media that produces hope and transformation in an age of crisis.

PACIFICATION POLICE UNITS

Every day young men from Rio's *favelas* are murdered, both by gang members and by the police (PERLMAN 2004). Much of this violence is highly gendered, racialized and classed, disproportionately affecting poor, African-Brazilian males (WACQUANT 2008; COSTA VARGAS and AMPARO ALVES 2010). Age also plays into this violence, as demonstrated by the renowned 1993 police execution of eight sleeping street children, known as the Candelaria massacre, which painted a particularly grim picture of one of the city's past attempts to sanitize its streets. Rio's violence is largely territorial-based, as drug gangs and police fight for control of the city's impoverished *favelas*. The *favelas* are over 900 in number, many of which ascend the city's steep slopes and provide limited access to essential services such as water and electricity. With one of the highest income gaps in the world, many of Rio's *favelados* are left with few options other than to eek out a living in the stigmatized urban informal sector. Others, particularly youth, turn to the more lucrative drug trade as a key strategy for household reproduction (WACQUANT 2008).

Mainstream media refer to Rio's *favelados* as *los marginais* – or 'the marginalized', a highly stigmatized word used to denote the "shiftless, dangerous ne'er do well" (PERLMAN 1977, 92) or alternatively, as home to *los bandidos* and their criminal accomplices. Media representations of *favelas* portray them as cesspits of violence harboring dangerous criminals, while protests against police violence are delegitimized by the presence of those associated with the drug trade. Meanwhile, women, children and elders are depicted as imprisoned in their homes, too fearful to leave lest they become victims of gang violence (see PERLMAN 2004). While some truth is found in these representations, particularly due to the reality of the city's extraordinary violence, few of these media representations capture the resilience of residents, nor their grassroots efforts, such as those demonstrated by *Visão Favela*, to organize against longstanding social, economic and political exclusion.

Santa Marta is one of Rio's oldest *favelas* and is located in the heart of the city, nudged up against some of the area's most valuable real estate. It is home to approximately 10,000 residents, housed in 1,000 to 2,000 hillside homes. Perched so close to Rio's beachside condos and key tourist spots, Santa Marta has become a prime target for 'pacification' and state-led gentrification. Along with the City of God, it has also become a symbol of the poverty and inequality suffered by Brazil's poor, most famously so in Michael Jackson's 1996 music video, 'They Don't Care About Us'. Lyrics from this video include: "I am the victim of police brutality, now; I'm tired of bein' the victim of hate; You're rapin' me of my pride; Oh, for God's sake I look to heaven to fulfill its prophecy; Set me free." Jackson's decision to film Santa Marta sparked great controversy among city authorities, as they declared the video would damage the city's ongoing efforts to rehabilitate its global image. Residents, however, welcomed the film crew in the hopes that Jackson's representations would bring greater awareness of their impoverished conditions (SCHEMO 1996).

Like many cities around the world, Rio is attempting to revamp its urban image in order to out-compete would-be urban contenders for both tourists and investors. Rhetoric promoting the perceived success of zero tolerance policies from New York City[1] has prompted many cities around the world to not only crack down on crime, but also 'regenerate', 'redevelop' and 'restructure' (SMITH 2002; LEE et al. 2008). SMITH (2002) argues that gentrification and revanchist urbanism – a punitive, right wing reaction against the poor in order to reclaim the city for the middle and upper classes – has become a global urban strategy connected to the spread of globalization and neoliberalism. While these types of policies have spread throughout the global north (MITCHELL 1997; BOWLING 1999; HERMER and MOSHER 2002; BELINA and HELMS 2003), they are increasingly gaining momentum across the global south, including in Shanghai (HE 2007), Mumbai (WHITEHEAD 2008), Santiago (LOPEZ-MORALES 2010), Mexico City (MOUNTZ and CURRAN 2009) and Guayaquil (SWANSON 2007). Yet, when punitive policies are implemented in the global south where socio-economic inequalities are severe, the spread of global urbanism is worrying. In Brazil the implementation of these

---

1   In the mid-1990s New York City mayor Rudolph Giuliani and Police Commissioner William Bratton launched a 'zero tolerance' campaign to crack down on the homeless, graffiti artists, squeegee kids, street youth and all those perceived as unsightly blemishes on the streets. These policies took inspiration from WILSON and KELLING's (1982) controversial Broken Windows theory, which suggests that intensive policing of low-level, anti-social behavior leads to an overall reduction in crime. This theory has since been widely critiqued (SMITH 1998; BOWLING 1999; HARCOURT 2001; MITCHELL 2003), yet it remains highly touted as a key municipal crime reduction strategy.

types of policies is particularly problematic due to pervasive and deeply entrenched inequality. WACQUANT (2003, 198) argues that New York-styled urban
policies "will have dramatic and far-reaching consequences on the social fabric of
state-society relations," particularly in Brazil, one of the champions of Giuliani's
zero tolerance policies.

Perhaps the most evocative representation of Rio's attempt to remarket itself
as an important global city was its successful bid to host the 2016 Olympic
Games. Leading up to the bid launch, Rio hosted the 2007 Pan-American Games,
an event widely viewed as a critical test run for the city's impending bid. While
mega-events can be used to mobilize resources for poorer neighborhoods (and are
often touted as opportunities to do so), the reality on the ground differs. As stated
by Raquel ROLNIK, a planning professor at the University of Sao Paulo in Brazil,
"unfortunately, the dominant approach we have seen with mega-events is that they
are part of the machinery of the territorial exclusion of the poor" (cited in RIOS
2010). As demonstrated with mega-events around the world, including the
Beijing, Vancouver and the forthcoming London Olympics, troubling social
problems are often erased from the public view in order to avoid the critical gaze
of an international audience. Instead a sanitized and cleansed image of the city is
packaged and delivered. Yet, the implementation of these cleansing strategies
varies in style; Rio's approach prior to the Pan-American Games was particularly
punitive and brutal.

In June 2007, a large-scale police operation was conducted in a series of
favelas on the city's northern periphery known as the *Complexo do Alemão*. In
what would later be dubbed the "*Complexo do Alemão* massacre," over 1,000
police officers invaded the community on June 27th and killed 19 civilians. Of the
deceased, 11 had no relation to the drug trade and the federal government later
conceded that several bore bullet wounds were consistent with execution-style
killings. Following a week of international headlines, condemnation from human
rights groups worldwide, and speculation that the invasion was related to a desire
to 'secure the city' prior to the Pan-American Games, Brazilian President Lula da
Silva visited the *Complexo do Alemão* and gave an impassioned speech declaring
state-*favela* relations would no longer be 'business as usual.' By this, da Silva was
referring to, on the one hand, the state's near total neglect in providing social
services, infrastructure, and basic necessities like running water, electricity, and
sanitation to many of the city's *favelas*. While on the other hand, the president
was also referencing over 25 years of an oft-criticized state security policy that
propelled periodic violent incursions into narco-controlled *favelas* by police
forces. Typical police invasions oftentimes consist of high intensity gun battles
that left untold numbers of innocent civilians wounded or dead. Once the police
operation was concluded, the unit traditionally departed the community thus ultimately leaving the traffickers once more in charge (WACQUANT 2008; COSTA
VARGAS and AMPARO ALVES 2010).

As da Silva spoke on the site that was the latest symbol of Rio's failed security policy, a new series of initiatives were introduced that would not only garner
more positive media attention going into the Pan American Games, but also

radically alter state-*favela* relations for years going forward. In terms of infrastructure and social services upgrades, da Silva pledged to direct R$1.2 billion from the federal government's *Programa de Acelerao e Crescimento –* Acceleration and Growth Program – (PAC) for *favela* upgrading. For da Silva, supplying basic services to communities in need was the best way to undercut the power of trafficking organizations: "If the state doesn't fulfill its role and does not provide services for the people," he said, "drug traffickers and organized crime will. We want people to have road access, street lighting, hospitals and schools" (BBC News 2007).

In addition to redressing the government's neglect in supplying infrastructure and social services, newly elected state governor Sergio Cabral and secretary of state security Dr. Jose Mariano Beltrame began crafting an alternative set of policies to counter the "invade-shoot-leave" approach that had brutalized the city's *favelados* over the last 25 years. Their solution was to create a series of community-based police units, consisting of younger, less battle-hardened, better paid, and better trained officers who were placed permanently in the *favelas*. Beltrame named his alternative police corps the *Unidades de Polícia Pacificadora* (UPP), or Pacification Police Units. Drawing parallels to the frontier mentality used by gentrifiers in New York City (SMITH 1996), the UPP was tasked to "pacify" the *favelados*, seemingly perceived as wild, savage and in need of taming. Theoretically, the UPP addressed several deficiencies of the preceding policy. First, an in-situ police force would both remove the presence of armed traffickers from the community and preclude the need for future violent police invasions. Second, a state security presence hypothetically allows the chronically neglected infrastructural improvements, social projects, and basic services to be implemented without interference.

It would not take long for the government to take action. In November of 2008 Special Forces invaded the *favela* of Santa Marta wresting control away from the local drug gang. One month later, the first UPP battalion was installed in the community. Between December of 2008 and June of 2010 a total of twelve *favelas*, with a combined population of well over 150,000 people, had UPP units in place. By this time, it was estimated that over 15 percent of the city's *favela* population was under UPP control (the government wants to double this figure by the end of the 2010). Authorities have publicly stated their desire to have UPP forces in each of Rio's more than 900 *favelas* by the time the Olympic torch arrives in 2016. The extent to which Rio's *favelas* will actually be 'pacified' by 2016 remains to be seen. The government's track record in comprehensively deploying infrastructure and social projects is poor. Historically, a small number of *favelas* have tended to receive the majority of NGO and government resources, thus contributing to deepening inequalities (CAVALCANTI 2007).

Seventeen months into the UPP initiative, it appears a calculus of risk and social visibility, informed by both damaging global headlines and the impending 2016 Olympic Games, is engendering the spatial distribution of Pacification Police units. As of June 2010, 'pacified communities' appear to fulfill one of the three following criteria. First, and most apparent, *favelas* lying adjacent to the wealthy neighborhoods and tourist districts of Rio's South Zone have been expressly targeted, including Santa Marta. Second, once the *favelas* proximate to the South Zone's wealthy areas had become thoroughly 'pacified,' the government shifted its focus to the middle-class neighborhood of *Tijuca*, in the North Zone. Coincidently, *Tijuca* also happens to be the site of several Olympic venues, including *Maracanã* Stadium, home of the opening and closing ceremonies. Finally, UPP forces have been placed in several of the city's more infamous *favelas*, at least in terms of media exposure, including, famously, City of God.

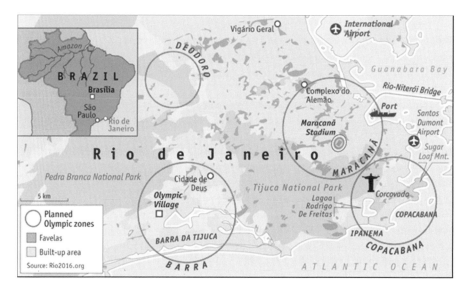

MEDIA FLOWS AND THE ENERVATED CITY

Although it is difficult to thoroughly assess a program still in its infancy, there has been one conspicuous and indisputable effect of Rio's UPP program: the media discourse emanating out from the city has changed. The international press corps assembled in Rio no longer sends stories about police massacres, extra-judicial killings, or the latest narco turf-battles back to their home bureaus in North America and Europe. None of the over 100 *favelas* still run by extra-governmental militias has received significant international press coverage in the last 12 months. One would also be hard pressed to find a recent story in an international daily documenting the city's extraordinarily high murder rates and/or the vast numbers of civilians Rio's police force continues to kill each year.

Instead, the global media have focused their collective gaze on Rio's new security initiative: the Pacification Police. Globally renowned media outlets such as *Newsweek* and The Economist have come to Rio in the last year to report on the UPPs. And by in large, they have liked what they've seen: "A magic moment for the City of God" reads the title of a June 2010 Economist article singing the praises of the UPP's presence in the City of God. They continue with a woman from the Cidade de Deus who states, "It was horror before…Now the children can play in the streets." While the article does include some skepticism regarding persistent police corruption, this is diminished as belonging to a "dreadlocked unemployed welder" and his friend, "who has the blank stare of a crack addict" (Economist 2010).

Mainstream media outlets have tended to position the efficacy of the UPP program in laudatory terms. As the vanguard of the UPP operation, Santa Marta has received the majority of this attention. A casual reading of press clippings on Santa Marta over the last 16 months would have the reader believe upwards of 95 percent of residents prefer the UPP to the drug traffickers: one year into the operation not a single resident has complained about police abuse or unwarranted searches. The reader might be led to believe the 'pacifying police' are indeed a kinder and gentler version of the notoriously corrupt, abusive and violent institution responsible for over 2,400 civilian deaths between 2007–2008; many of those killings thought to be extra-judicial according to Human Rights Watch. "Rio police show new face in battle-hardened slums" reads the title from a *Reuters* January 2009 article. The media have also been keen to voice the opinion of government authorities, like Rio state governor Sergio Cabral who recently told the New York Times: "The testimonials I have received from the people that have been freed from the parallel power are just incredible. We are now free from terrorism. Finally, governor, I can sleep at night (New York Times 2010)." Testimonials from actual Santa Marta residents, however, have been few and far between in the major international and domestic dailies. From the outside looking in, then, the Unidades da Policia Pacificadora appear to be the solution to the city's now infamous image problem of narco-territorialized slums, uncontrolled violence, and egregious levels of police abuse.

Yet, the same casual reader of Santa Marta press clippings might have been quite surprised if they were in the community in March of 2010, when a large rally was held to contest the illegality of UPP policing practices. Representatives from Humans Rights Watch, the Public Defender's office, and the Residents Association were in attendance to witness over 2,000 pamphlets passed out to community residents detailing how to resist police misconduct, illegal searches, and other human rights violations. The pamphlets, put together by Santa Marta's grassroots community organization, *Visão Favela*, in coordination with local lawyers and human rights groups, was the product of 15 months of perceived excessive UPP police abuse. Youths were illegally arrested for not carrying identification cards. Houses were illegally searched. A few residents, primarily adolescent males, were stopped and searched as many as 10 times in one day. For one Santa Marta resident, the UPP were no different than the notoriously brutal Military Police: "it's the same truculent police that walked up the hill in the past, the same police who abused the residents. It's the same, no differentiation at all" (Cultura NI, April 2010).

In addition to the excesses of 'community policing', Santa Marta residents are also subjected to an acute form of social and cultural displacement. Soon after the

UPP consolidated in the community they banned hip-hop performances and the extremely popular community dance parties known as *Baile Funks*. The *Baile Funk* is a *carioca* cultural phenomenon both born and practiced almost exclusively in Rio's *favelas*. It is a vibrant youth culture of no small significance to *favelados*, as evidenced by the following description of the *baile* by its principal ethnographer Paul SNEED (2008, 60): "At its core, funk music is a transformational and countercultural practice through which these young people experience a sense of unity and find a greater sense of courage to resist the wearying effects of the harsh realities they face on a daily basis." The UPP, however, claims *Baile Funks* are little more than venues for drug distribution and they are therein justified in prohibiting this vibrant musical practice. *Pagode*, *Samba*, and other 'traditional' forms of musical expression enjoyed mostly by older members of the community are still allowed. Meanwhile, youth are left with few outlets for diversion.

The UPPs regulate other social events in the community. Parties, festivals, and cultural events must now be negotiated with the police. Itimar Silva, founder of Santa Marta's youth outreach program *Grupo Eco*, wrote an open letter to Santa Marta residents in June of 2010 criticizing the UPPs suppression of cultural activities and other 'festive events' beyond hip-hop and funk:

> It is increasingly clear that their goal is the territorial control of the favelas and the imposition of a behavioral pattern defined as good and right by the police. So if the residents behave "properly" according the standards of the police, they have access to the benefits (public policies) offered by the state and implemented from the UPP, and, perhaps, celebrations, within the limits predefined by the police (Visao Favela, June 2010).

In imposing control measures such as these, the UPP crushes desire and passion of Santa Marta's youthful residents.

LEFEBVRE's spontaneous and corporeal city contains the right to urban life, but authoritarian power regards spontaneity as the enemy and so, by crushing youthful desire and passion from Santa Marta's urban life, then so too a vital spark and the possibility of the political is removed. When *Visão Favela* published on its website a picture of the smiling and youthful Chief of Police for Santa Marta's UPP holding a 2016 Olympic flag before an iconic Rio backdrop it did so as an ironic gesture with the intent of creating media turbulence.

Residents of Santa Marta also complain about a phenomenon known locally as *remoção branca*, or white removal. *Remoção branca* is a guised form of economic displacement in which the community is suddenly subjected to an unceasing wave of new expenditures and rent increases beyond the means of many long-term residents. According to Santa Marta resident MC Fiell, the founder of *Visão Favela*, since the UPP arrived in Santa Marta in December of 2008 a typical apartment, which would have rented for about $150 *reais* prior to the UPP, now costs upwards of $400 *reais*. The UPP also abruptly shut down Santa Marta's previously free access to electricity and cable television. Residents are now forced to pay what community residents complain is a widely fluctuating and expensive electricity bill. The majority of Santa Marta residents simply do not have the disposable income necessary to absorb an extra $300 to $400 *reais* in expenditures per month. Many residents either earn near the nation's minimum wage of $465 *reais* working service sector jobs or they are employed in the increasingly volatile informal sector. *Visão Favela* estimates at the very bare minimum a single person would need $570 a month to live in Santa Marta. Those unable to meet the new economic demands are either displaced from the community or forced to move in with neighbors or family members. For Fiell, *remoção branca* is nothing more than a slightly more nuanced version of the same *favela* displacement policies the Brazilian government has used regularly over the last 100 years:

> Politicians are changing their theories of policy. Previously, they spoke to, literally, removal of slums. Today the policy is not to say that we will remove, but they are going to do is make things more expensive…The policy goes, though not literally, moving people from Santa Marta (Cultura NI, April 2010).

It is not unreasonable to imagine that a large percentage of renters in the community may be forced to move if current rent and utility prices remain constant. Thus, slum removal policies have shifted to more insidious gentrification strategies (SLATER 2008).

In each of the *favelas* occupied by the UPP since 2008, the government has also promised 'social occupation' consisting of significant government investments in social services such as health clinics, schools, and community centers. In Santa Marta, government funds have indeed been distributed to a number of projects. For example, a new soccer pitch and a transportation cable linking the bottom of the hill to the top were built. Yet, a large amount of government expenditures poured into the community between December of 2008 and June of 2010 have been geared towards 'security measures'. For instance, some three million *reais* were spent to build a large Gaza-esque wall around the perimeter of the community. Although portrayed as an eco-barrier to prevent Santa Marta from expanding further into the surrounding Atlantic rainforest, residents perceive it as an attempt to segregate and hide the slum from adjacent beachfront condos, particularly in light of the forthcoming Olympic Games (Darlington 2009). Along this newly-built wall, a graffiti artist asks: *muro para quem,* wall for whom? (see CALDEIRA 2001). An additional half million *reais* went into installing surveillance cameras throughout the *favela*. As of June 2010 no new schools, health clinics, or cultural centers have been built. The government's priorities could not be clearer: to ensure the police occupation and pacification of the community prior to rolling out any type of 'social occupation.'

The government's and mainstream media's complete lack of interest in either addressing or highlighting the threat of displacement in Santa Marta suggests the *status quo* will continue unless community-based interventions can successfully resist the de-localizing and de-corporealizing effects of state intervention. In the following section we discuss attempts to resist displacement by constructing an alternative politics of place based on articulations of difference, emergence, and with respect to local cultural and economic practices. Particularly important for *Visão Favela* construction(s) of Santa Marta as a 'figured world' (ESCOBAR 2008), has been a series of activist media projects that the group distributes through its various networks.

MEDIA TURBULENCE AND THE EBULENT CITY

> The point of Visão Favela is to show our reality to the rest of the world. Because, formerly, it's always been people from the outside that come here and make things up about the inside (Fiell, Personal Interview, June 2010).

ESCOBAR (2008, 65) argues that strategies for battling displacement

> should take as a point of departure an understanding of resisting, returning, and re-placing that is contextual with respect to local practices, building on movements for identity, territory, and autonomy wherever they may exist.

More specifically, he is interested in the ability of activist groups to create 'figured worlds' in which those local practices, culture, and identities are deployed effectively enough to create a visible (spontaneous, emotional and corporeal) space for authoring which may contest the current dominant representation(s) of that place. Or, to put it differently: to create turbulence in WILLIAMS' irresponsible flows of images and feelings.

Visão Favela – the 'vision of the *favela*' – was launched in Santa Marta in 2007 as a platform for residents to express their views of life in the community. The initial post on their activist weblog was titled: *a vision that many do not see*. By this, the organization was referring to the everyday cultural and economic practices, feelings and images that reproduce life in Santa Marta, yet are rarely included in mainstream representations of the *favela*.

"If you do anything with culture in the *favela*, the media doesn't want anything to do with it," explains Fiell (*Personal Interview*, June 2010). To redress this imbalance, *Visão Favela* works towards circulating images of the community's indigenous cultural activities – such as hip-hop, *capoeira*, art, graffiti, etc. – through information and communication technologies (ICTs), cinema, and traditional journalism.

For an organization that receives no government or NGO financial aid, *Visão Favela* has been quite successful at producing a high volume of grassroots media projects. In 2008 they completed a short documentary, *788* (a reference to the number of steps it takes to walk from the base of Santa Marta to the top), emphasizing the social realities of living in Santa Marta. The film, which Fiell describes as representing the "total geography of the *favela*, from those that live up high to those that live at the bottom," was screened in Holland where it won several awards. A new short film, *Winged Kite*, is in production. Several hip-hop albums and music videos have been independently produced; the songs use consciousness-raising lyrics to discuss both the productive capabilities of *favelados* and the various forms of discrimination they suffer. The organization also distributes a free community newsletter/fanzine showcasing the artistic talents of Santa Marta youth. The fanzine was "born of the need to reproduce the truth" about *favelas* and is dedicated to "sharpening the minds of young people and adults" (Visao Favela, April 2010). In addition to these media projects, *Visão Favela* currently sponsors a poetry project, study groups, a web-based television channel, and a community library entitled 'Evolution.' Each of these projects is electronically linked and accessible through the organization's central network node, and an activist blog www.visaodafavelabr.blogspot.com, thus making materials readily available for public consumption. *Visão Favela's* grassroots media blitz is productive in the sense that it works towards assembling an alternative visuality of the *favela*. It engages in a struggle over the construction of place identity.

Echoing WILLIAM's 'mobile privatization', ESCOBAR (2008, 32) approaches conflicts over the production of locality as being tantamount to two conflicting, yet at times mutually constitutive 'processes of localization'. On the one hand, there are the dominant forces of the state, capital, and mainstream media which attempt to "shift the production of locality in their favor," thus ultimately creating "…a delocalizing effect with respect to places." In the case of Rio's *favelas*, dominant flows of media representations, from the cottage industry of '*favela* gangster films' to the news reports which only write about *favelas* when they turn violent, flatten the meaning of the '*favela*' and exported it to such an extent that each of Rio's more than 900 *favelas* do not have identities of their own. Rather, they are each the *favela* of *City of God*, bounded war-zones filled with trigger happy adolescents, drugs, and inescapable poverty. Place, de-localized. Prior to the UPP's 2008 arrival in Santa Marta, this domineering form of localization, scaled-down onto each of Rio's *favelas,* was the very thing *Visão Favela* sought to resist.

One informative and paradigmatic example of *Visão Favela's* resistance is the music video for the *788* rap song. The video's initial scenes show a series of zoomed-out, panoramic images of Santa Marta's exterior that are reminiscent of how the community is most often gazed upon by the mainstream press: from the outside peering in. The following shot is then filmed from the vantage point of a helicopter, again invoking the frequently consumed 'birds eye' visuality of the *favela* most commonly used by TV news stations when they cover flair-ups of violence in community. This is LEFEBVRE's *representation of space* (1991, 50)

"…in thrall to both knowledge and power…adolescence perforce suffers from it, for it cannot discern its own reality therein: it furnishes no male or female images nor any images of possible pleasure." Soon thereafter, however, the camera is thrust into the interior of the *favela* where a slew of cultural activities are occurring. This shift then is to LEFEBVRE's *representational space,* which is embodied, sensual and intimate. Youth are shown in the public spaces of the community making art, practicing hip-hop dancing, and playing *capoeira*. Various sorts of musical performances are taking place in the *becos* (alleyways) and *pracas* (squares) of the *favela*. Young children, for example, are drumming samba beats, adolescents are playing the guitar and bass, and a twenty-something male is playing the *berimbau* (a string instrument used in *capoeira*). A large group of residents are shown encircling a break-dancing crew as they perform in one of the community's larger public spaces. And, finally, Fiell and his crew are found rapping at the summit of the *favela*. In all, over 30 actual residents of Santa Marta are shown in the six minute video clip, and none of them are using or selling drugs, carrying a weapon, or appear to be related to organized crime in any sense. Rather, they are living with passion and spontaneity, while children are playing with impunity in the streets, all of this is contrary to the image the mainstream media presents.

The *favela* of *788* is reminiscent of the corporealized city of Henri LEFEBVRE: a self-built community with abundant public spaces for fulfilling the bodily needs and desires of its residents; yet still untouched by planning professionals, architects, and the state. The video invokes the pleasures of the community through highlighting its everyday cultural practices. Critically, the desires of *788's* protagonists are realized as they are (in)placed within the physical space of the *favela*; its *becos*, *pracas*, and *quadras*. Their bodies find pleasure in these self-built community spaces, free from either territorializing, drug gangs or morally regulating pacification units. "In as much as adolescents are unable to challenge either the dominant system's imperious architecture or its deployment of signs," writes LEFEBVRE (2001, 50), "it is only by way of revolt that they have any prospect of recovering the world of differences – the natural, the sensory/sensual, sexuality and pleasure."

The arrival of the UPP engendered new and unique sets of dislocating media images, over-coded with sterile abstractions of youthful, Olympic vigor. The mainstream media presented post-UPP Santa Marta as a safe, idealized model-*favela* full of pacified and appreciative *favelados*. Laudatory television reports about the community almost always use the evocative image of one particular section of the *favela*, *Praca Cantao*, which was recently painted in an array of tasteful colors by two Dutch painters.

For Fiell, the circulation of these gentrified images, which he refers to as "the beautiful movie that TV Globo loves to show about Santa Marta (*Personal Interview* June 2010)," is one reason wealthier outsiders are increasingly interested in coming to Santa Marta to purchase housing and thus elevating property values. As the effects of the UPP in Santa Marta and other communities became more apparent, *Visão Favela* began devoting an ever increasing share of its media space to raising consciousness about this new reality.

First, *Visão Favela* began expanding its activist networks in an attempt to more thoroughly cut through the media's tacit endorsement of the UPPs. The group has been particularly successful at creating both vertical and horizontal links with other networks. For instance, on the one hand, they have created vertical networks with law firms, NGO's, and Human Rights organizations in order to both ensure that the constitutional rights of Santa Marta's citizens are upheld and to create solid links with institutions capable of airing the communities' grievances to government officials. Partnerships have been formed with Amnesty International, Global Justice Brazil, and The Center for the Defense of Human Rights amongst others. On the other hand, *Visão Favela* attempts to export their localized knowledge(s) on how to fight against displacement (economic and cultural) and police abuse to other communities facing similar circumstances. For example, their *Cartilha Popular do Santa Marta*, a 32 page booklet vividly illustrating citizen's constitutional rights as they relate to illegal police actions, has been distributed to a number of other *favelas* in Rio de Janeiro and beyond. The group has also gone on several lecturing tours at local colleges to raise consciousness about the UPPs. These events have typically consisted of a film screening and lecture, followed by a brief hip-hop performance.

Second, *Visão Favela's* blogspot has become the *de-facto* site of UPP resistance journalism. The organization works hard to fill in the absences left by mainstream media's venerating depiction of 'pacified *favelas*'. Residents from

Santa Marta and other UPP occupied *favelas*, for example, can be found on the website giving testimonials about conditions in their communities via podcasts. "The truth about the UPPs on the Morro da Providencia" is the title of one such podcast given by the president of the residents' association of Morro da Providencia, a recently pacified *favela*. Any instances of abuse suffered at the hands of the Pacification Police are also reported. For instance, an article was recently written about the strip search and subsequent detention of famous Brazilian film director Rodrigo Felham in the City of God. Felham, the story reports, was leaving his house for Cannes, France to screen his latest film when the police accused him of carrying illegal drugs or guns and demanded he pull his pants down in the presence of other residents of the community. Finally, members of *Visão Favela* write consciousness-raising essays discussing a range of topics. Themes have included everything from racial and class discrimination, to the connection between (direct) *favela* removal policies of the past and the contemporary context of *remoção branca*.

THE FIGHT TO STAY PUT

With this essay, we demonstrate the ways that civic authorities close down and pacify Rio's *favela* streets in the wake of a mainstream media blitzkrieg that was touched off by the city's desire to attract the 2016 Olympics. The image event of violent *favelados* as famously portrayed in *City of God* is used as a reason to create further spectacular images and policies as foils to the vibrant, sensual, corporeal practices of everyday *favela* life. The policies that created the Pacification Police Units are successful to the extent that the *favela* streets are patrolled and controlled, and local spontaneity and passion is dampened. Challenging these dominant forms of localization and emerging from the epistemic borders of place-based conflicts, are what ESCOBAR (2008, 32) refers to as subaltern forms of localization: "place-based strategies that rely on the attachment to territory and culture." *Visão Favela's* grassroots media activism, then, can be understood as a series of subaltern acts attempting to construct an alternative visuality of their *favela* against the over-coded '*favela*' of *City of God*. In this sense, LEFEBVRE's (1991) *representations of space* take on a different twist as mainstream media use a particular form of coding to create spectacle, and policy makers use the same coding to justify the repressive tactics of the UPP.

For LEFEBVRE, the corporeal city is a spontaneous city. More than likely, LEFEBVRE would have sided with TUSHNET's (1984) critique of rights-based liberal theories in favor of more immediate action. In the face of oppressive UPP measures, *Visão Favela* provides an immediate reaction. It flares, amongst other things, a spontaneous turbulence in the flow of sanitized corporate media and this has the capacity of a prodigiously creative force. "Power", wrote LEFEBVRE (1996, 70), "regards spontaneity as the enemy." LEFEBVRE's Marxist politics saw the street as the public place where spontaneous resistance, contestation and revolution occur to transform everyday life. Powerful institutions fear the street, and create forces like the UPP to cordon it off, and dampen its passions. Streets are explicitly political, but they are also localized, and a symbol of only "partial practice." Andy MERRIFIELD (2002, 87) notes that if spontaneous street-contestation is the strength of political uprising, then it is also its weakness because streets can be controlled by powerful institutions. Media has the capacity to fill this political void, providing a fuller practice that is mobile, that is hard to cordon off and pacify; a practice that creates turbulence in the flow of irresponsible global media.

In the face of UPP's attempted enervation of Santa Marta's youth culture, *Visão Favela's* successful disruption of mainstream media – 'to set the story right' – creates new forms of youth passion and desire. Pacification of *favela* streets is followed by a dangerous ebullience from young people through the media of *Visão Favela*. And because it is difficult to cordon-off these new media, there is leakage to other *favelas* and other young people. In so doing, *Visão Favela* embraces Escobar's desire for place-based *redes* that are ill-fitting and constantly in need of repair but are also indomitable. *Visão Favela* creates that immediacy and a desire to stay put that is based on corpulent local passions rather than contrived and anesthetic Olympic aspirations.

REFERENCES

BBC News (July 2007): Brazil Launches Slum Reform Drive. Jul 3rd, 2007. Accessed January 12, 2008. http://news.bbc.co.uk/go/pr/fr/-/2/hi/americas/6263750.stm.

BELINA, B and G. HELMS (2003): Zero tolerance for the industrial past and other threats: Policing and urban entrepreneurialism in Britain and Germany. *Urban Studies.* 40: 1845–1867.

BENJAMIN, W. (2008): *The Work of Art in the Age of its Technological Reproducibility and Other Writings on Media.* Harvard.

BOWLING, B. (1999): The rise and fall of New York Murder. Zero Tolerance or Crack's Decline? *British Journal of Criminology.* 39: 531–554.

CALDEIRA, T. (2001): *City of Walls. Crime, Segregation, and Citizenship in São Paulo.* Berkeley.

CASTREE, N., P.; CHATTERTON, N., W. HEYNEN, W. LARNER and M. W. WRIGHT (Eds.) (2010): *The Point is to Change it: Geographies of Hope and Survival in an Age of Crisis.* Oxford.

CAVALCANTI, M. (2007): *Of Shacks, Houses and Fortresses: An Ethnography of Favela Consolidation in Rio de Janeiro.* Phd Dissertation. The University of Chicago, Department of Anthropology.

COSTA VARGAS, J. and J. AMPARO ALVES (2010): Geographies of death: an intersectional analysis of police lethality and the racialized regimes of citizenship in São Paulo. Ethnic and Racial Studies. 33: 611–636.

Cultura NI (2010): Algo de Podre no Reino das UPPs. April 11th, 2010. Accessed April 29th, 2010. http://culturani.blogspot.com/2010/04/o-hip-hop-do-santa-marta.html.

DELEUZE, G. (1986): *Cinema 1: The Movement-Image.* Trans., Hugh Tomlinson and Barbara Habberjam. London.

ESCOBAR, A. (2008): *Territories of Difference: Place, Movements, Life,* Redes. Durham and London.

HARCOURT, B. (2001): *Illusion of Order: The False Promise of Broken Windows Policing*. Cambridge.

HAYS, J. (2009): Departures: The 21st Century Hotel – Your Media/Home Away From Home. CLARKE, D. B., V. CRAWFORD PFANNHAUSER and M. A. DOEL (Eds.): *Moving Pictures/Stopping Places.* Plymouth, UK: 371–380

HE, S. (2007): State-sponsored gentrification under market transition: the case of Shanghai. *Urban Affairs.* 43: 171–198.

HERMER, J. and J. MOSHER (Eds.) (2002): *Disorderly People. Law and the Politics of Exclusion in Ontario.* Halifax.

HUBBARD, P. (2006): *The City*. New York.

LEE, L., T. SLATER and E. WYLY (2008): *Gentrification.* New York.

LEFEBVRE, H. (1991): *Writing on Cities.* Translated by E. Kofman and E. Lebas. Oxford.

LOPEZ-MORALES, E. J. (2010): Real estate market, state-entrepreneurialism and urban policy in the 'gentrification by ground rent dispossession' of Santiago de Chile. *Journal of Latin American Geography.* 9: 145–173.

LOW, S. (2000): *On the Plaza: The Politics of Public Space and Culture.* Austin.

MERRIFIELD, A. (2002): *Metromarxism: A Marxist tale of the City*. New York and London.

MITCHELL, D. (1997): The annihilation of space by law: the roots and implications of anti-homeless laws in the United States. *Antipode.* 29: 303–335.

MITCHELL, D. (2003): *The Right to the City: Social Justice and the Fight for Public Space*. New York.

MOUNTZ, A. and W. CURRAN (2009): Policing in drag: Giuliani goes global with the illusion of control. *Geoforum.* 40: 1033–1040.

New York Times (2010): With World Watching, Rio Focuses on Security. January, 16th 2010.

PERLMAN, J. (1977): *The Myth of Marginality: Urban Poverty and Politics in Rio de Janeiro.* Berkeley.

PERLMAN, J. (2004): Marginality: from myth to reality in the *favelas* of Rio de Janeiro, 1969– 2002. Pages 105–146 in ROY, A. and N. ALSAYYAD (Eds.): *Urban Informality: Transnational Perspectives from the Middle East, Latin America, and South Asia*. Oxford.

RIOS, K. (2010): Brazilian crackdown: It's Giuliani-time as Rio de Janeiro goes for the gold. *The Indypendent.* March 31 2010. http://www.indypendent.org/2010/03/31/brazilian-crackdown/ (accessed June 15, 2010).

SCHEMO, D. J. (1996): Rio frets as Michael Jackson plans to film slum. *The New York Times.* February 11 1996. http://www.nytimes.com/1996/02/11/world/rio-frets-as-michael-jackson-plans-to-film-slum.html?pagewanted=1 (accessed June 24, 2010).

SLATER, T. (2006): The eviction of critical perspectives from gentrification research. *International Journal of Urban and Regional Research.* 30: 737–757.

SMITH, N. (1998): Giuliani time: The revanchist 1990s. *Social Text.* 57: 1–20.

SMITH, N. (2002): New globalism, new urbanism: Gentrification as global urban strategy. *Antipode.* 34: 428–450.

SNEED, P. (2008): Favela Utopias: The *Bailes Funk* in Rio's Crisis of Social Exclusion and Violence. *Latin American Research Review.* 43: 57–79.

SWANSON, K. (2007): Revanchist urbanism heads south: the regulation of indigenous street vendors and beggars in Ecuador. *Antipode.* 39: 708–728.

The Economist (2010): A magic moment for the City of God. *The Economist.* June 10, 2010. http://www.economist.com/node/16326428 (accessed June 25, 2010).

TUSHNET, M. (1984): Essay on Rights. Part of a Symposium on A Critique of Rights. *Texs law Review* 62, 1363–1412.

Visao Favela (2010): http://visaodafavelabr.blogspot.com/ (accessed April 25, 2010).

WACQUANT, L. (2008): The militarization of urban marginality: lessons from the Brazilian metropolis. *International Political Sociology.* 2: 56–74.

WACQUANT, L. (2003): Towards a dictatorship over the poor? Notes on the penalization of poverty
    in Brazil. *Punishment & Society.* 5: 197–205.
WILLIAMS, R. (1974/1992): *Television; Technological and Cultural Form.* Hanover.
WILSON, J. and G. KELLING (1982): Broken windows. *The Atlantic Monthly.* March: 1–10.
WHITEHEAD, J. (2008): Rent gaps, revanchism and regimes of accumulation in Mumbai.
    Anthropologica. 50: 269–282.

Amy Siciliano / Paul S. B. Jackson

'MORE WESTCHESTER THAN WATTS': LEARNING TO LOVE THE CITY THROUGH *SESAME STREET*

> Come and play
>
> Everything's A-OK
>
> Friendly neighbors there
>
> That's where we meet
>
> (*Sesame Street* Theme "Sunny Day", Raposo, Stone and Hart, 1970)

[T]he imagination of the child is conceived as the past and future utopia of the adult. But set up as an inner realm of fantasy, this model of his Origin and his Ideal Future Society lends itself to the free assimilation of all his woes. It enables the adult to partake of his own demons, provided they have been coated in the syrup of paradise, and that they travel there with the passport of innocence (DORFMAN and MATTELART 1975, 31).

In this chapter we explore the role of *Sesame Street* as an American cultural response to urban crises of 1960s. Specifically we look at how the urban fantasy-land projected through this children's educational television program helped re-shape meanings associated with the city and urban social identities. We suggest that *Sesame Street's* sanitized 'ghetto' aesthetic and sterilized social scripts helped to generate a new image of the city as a desirable locale for the white middle classes. In projecting an idealized New York streetscape into the homes of millions of Americans it helped alter public discourses and representations about urban space and middle-class identification with these spaces.

While progressive in its intent, *Sesame Street* nevertheless muted conditions generating highly spatialized structural inequalities in postwar America while simultaneously presenting an image of the city fit for bourgeois consumption. While the show's producers strived to raise educational levels of poor inner city children, they did so primarily through the ideals of white liberal America. In projecting an aestheticized vision of urban life from an image of their own making, they disassociated urban social problems from the structural inequalities of post-war America, embedding them instead in street-level praxis. We conclude by suggesting that in giving white suburbanites an image and narrative of the inner city far 'more Westchester than Watts',[1] *Sesame Street* should be placed alongside other cultural icons of the era like Jane JACOBS (1961). Both – however inadvertently – contributed to inner city gentrification and displacement by helping white middle class suburbanites learn to love the city.

Our purpose for this chapter is not to add to the various popular histories of the production of the *Sesame Street*, written mostly from the viewpoint of educational television. Rather our contribution re-reads the histories of children's educational television within the context of L.B Johnson's War on Poverty to show how the cultural production of this program helped to construct a pastoral version of urban life that intersected with the cusp of a 'new orthodoxy...hailing urban regeneration and neighbourhood revitalization' as strategies to save the city from crisis (PALEN and LONDON 1984, 4; SMITH and WILLIAMS, 1986). Such strategies depended on promoting an image of the city far removed from the reality of everyday life for the urban poor, many of whom could not afford, or were not allowed to flee declining city centres. Sesame Street was part of the historical processes normalizing a liberal image of urban life. While today for instance, a commonplace image of urban streetscapes is one constitutive of culture, creativity, and authentic community (see for instance FLORIDA 2002), when an inner city street was proposed as the set for Sesame Street, the suggestion was received as shockingly risky due to its association with violence, decay and abandonment.

RESPONDING TO URBAN CRISES: IMAGES, EXPERTS AND EDUCATION

In the immediate post-war era, the key issues facing America's major urban centres came to be understood and defined in terms of physical and technological problems – housing conditions, traffic congestion, economic decentralization and suburbanization. Planners and bureaucrats aimed to remedy the 'blight' affecting U.S cities through grand urban renewal schemes such as 'slum' eradication, highway and high-rise construction. The massive upheaval such renewal projects bore for many existing urban residents prompted grassroots response – most notably Jane JACOBS's pioneering work *The Death and Life of Great American Cities* (1961).

Death and Life surfaced at a time when many cities were in the midst of decline as a consequence of abandonment and disinvestment by both labour and capital in the immediate post-war era. As many scholars have documented JACOBS helped shape the transition of public discourse over the health and fate of American cities by bringing to light an alternative vision of economic, social and cultural opportunities afforded by an urban lifestyle (NICOLAIDES 2007). Perhaps most profoundly was JACOBS's argument that a good urban neighbourhood

---

1   The phrase comes from Joan Ganz COONEY, one the key producers of the show. MORROW, R.
    W. 2006. *Sesame Street and the reform of children's television.* Baltimore: Johns Hopkins
    University Press. p. 153.

operates as a 'street ballet', where diverse social groups come together in organic, messy, spontaneous, and serendipitous ways. This active ballet of people and commerce gave cities their vitality, charm and character, kept them safe from crime, and most importantly offered the greatest hope civility and progress among humankind. Over time, her ideas – targeted not only to planners, but to the white middle classes who had fled the inner city for the suburbs – would come to play an instrumental role in normalizing an alternative narrative of what urban life could be, one that while seemingly radical at the time, was soundly rooted in a white, liberal perspective (BERMAN 1982). It is perhaps unsurprising then to see how JACOBS's partial and selective vision of an inner city neighbourhood came to occupy such a privileged position in popular and planning circles: as Peter Drier notes:

> [P]erhaps more than anyone else during the past half century, JACOBS changed the way we think about liveable cities. Indeed, it is a mark of her impact that many people influenced by her ideas have never heard of her. Her views have become part of the conventional wisdom, if not always part of the continuing practice, of city planning (DREIER 2006).

In the coming decades, the broad circulation of JACOBS's aestheticized imaginary of urban life as the site of 'authentic community' combined with the increased association of the suburbs with various social pathologies (an association JACOBS played a central part in generating) encouraged young professionals to seek out inner city neighbourhoods as places to call home. These sites, as desirable locales for the mobile middle and upper classes, contributed to the displacement and/or criminalization of poor, racialized and working class people (DAVIS 1990; SMITH 1996; LEY 1996).

Shortly after the publication of JACOBS's book, a series of urban uprisings cut through Chicago, Philadelphia and other smaller cities in eastern and southern states. These would continue throughout the decade, stretching across the U.S, in 257 cities. In the south they were largely a response to hostile reactions to desegregation legislation, and in the North and West to chronic unemployment, poor housing conditions and the continuation of racial discrimination in housing and labour markets. As SUGRUE (1996) demonstrates in his study of the urban crisis in Detroit during this period, urban unrest occurred during a prolonged period of declining wages and chronic under and unemployment particularly among African Americans, many of whom were forced to remain in declining urban centres due to discriminatory suburban housing polices and lending practices.

By the mid-1960s, the growth in extreme poverty within the United States was brought to public attention due to influential books like Michael HARRINGTON's *The Other America* (1963). At the same time a series of governmental commissions responding to the urban uprisings helped reposition the root causes of urban decline in *social* rather than physical or technological terms.[2] Yet even as one of the most widely publicized of these reports pinpointed white racism and the structures that supported it as a primary cause of urban decline and social unrest (KERNER 1967), the "geographic focal pointmoved...to the ghetto" (BEAUREGARD 2003, 131). In other words, the origin of urban decline and the social crises associated with it crystallized in the social space of the "ghetto", remaining conceptually distinct from the spaces and institutions of white privilege.

The ghetto as the primary symptom of America's social problems would profoundly shape the form and content of President Lyndon B. Johnson's vision and action plan for social reform. His liberal vision of the Great Society offered Americans hope to overcome extreme inequality in the U.S, through the War on Poverty, the Civil Rights Act of 1964, and the Voting Rights Act of 1965. Under Johnson's administration, experts turned their sights on strategies to control violence and maintain social stability within the framework, rhetoric, and dreams of liberal America.

One key program that directly came from the War on Poverty was Head Start. Head Start focused on preschool education for the poor. Early childhood education would become an instrument of social reformbuilt upon extensive social science research. Educational television also was seen as a key strategy to shrink the divide in education levels between the "disadvantaged" and the middle-class; a revolutionary new medium that was able promote social change. Television was emerging as a nationally shared medium and the airwaves, as a national resource. At the time cable television technology was being promoted as a means to remedy alienated "ghetto dwellers". In particular hope rested in televised media as a place where various cultural producers could respond to social divides prevalent in American society. America's poor was isolated from society, and cable technology became the infrastructure that could link communities together, reduce citizen alienation, and improve relations between citizens and government (LIGHT 2003, 171). Cable's first decade, (1966–1976) coincided with the rise of *Sesame Street*. While *Sesame Street* was not initially shown on cable, taken together television was talked about as a technological fix to social and economic problems. By 1972 the Federal Communications Commission (FCC) mandated channels for citizen programming, municipal information, and educational content. The Great Society would be broadcast and piped into every American home. *Sesame Street* was part of that programming. As one critic of the show remarked:

> What most closely ties *Sesame Street* to the Great Society is the way it served different social classes and cultural constituencies. Its was a program for discontented middle-class parents, but also made blacks and Hispanics visible on children's television" (MORROW 2006, 35).

To take childhood entertainment seriously as a cultural and ideological force, we rely on Ariel DORFMAN and Armand MATTELART's book titled *How to Read*

---

2 See for instance, a series of governmental commissions on discrimination, unemployment and social unrest such as MCCONE (1965), MOYNIHAN (1965), KERNER (1967), and SCRANTON (1970) (for a history of these commissions see GRAHAM 1980).

Donald Duck, published in 1971. The authors claim that children's media is "no mere refuge in the area of occasional entertainment; it is our everyday stuff of social oppression" (DORFMAN and MATTELART 1975, 98). While *Sesame Street* is media directed towards children, the content says more about the producers than the audience. The show reflected the producers' "political utopia in the imaginative world of the child" (DORFMAN and MATTELART 1975, 59). White producers create a televised imagination that impoverished African America children could identify with. Yet as *Sesame Street* became a national show the relations between the constructed ghetto street utopia (a "black" setting) was then broadcasted into the domestic spaces of the American homes (a "white" setting).[3] *Sesame Street* domesticated racial conflict and urban poverty through a localized and sanitized aesthetic and praxis, and projected this view into living rooms across the United States. In doing so, it transformed this feared social space into something familiar and safe; the structural working of "social oppression", reified into personally resolvable "everyday stuff".

THE ROLE OF CULTURAL PRODUCERS

Sesame Street was conceived at a 1966 dinner party hosted by a TV producer Joan Ganz COONEY in her midtown Manhattan apartment. COONEY, who had already won an Emmy from a television show on urban poverty, invited over to dinner Lloyd Morrisett, Vice-President of Carnegie Corporation, and her boss Lewis Freedman who ran Channel 13, New York City's educational television channel. As COONEY served beef bourguignon, the topic of conversation turned to television. Morrisett raised a question to the group, "Do you think television could teach children?" Morrisett drew on seed funding from the Carnegie Foundation's new mandate to tackle poverty in the United States through social science research (LAGEMANN 1989, 216), to hire COONEY as a television consultant. The perspective of the 'detached intellectual' would shape their agenda: according to Morrisett "we would term ourselves progressive…Our support [of civil rights and social justice] was more intellectual than action oriented" (DAVIS 2008, 12). Specifically, the creators of *Sesame Street* aimed to respond to the struggles and turmoil of the 1960s through their interventions in the realm of culture and the media.

What piqued researchers interest in the potential of television to raise educational levels was the well documented observation that children could repeat commercial advertisements with ease. The question was could such learning behaviour be harnessed to educate preschoolers? COONEY spent three months interviewing and preparing a report called *The Potential Uses of Television in Preschool Education*. It stated that the prevailing view in childhood cognitive development was that aggressive intervention in early education could com-

---

3   For a summary of the gendered and racial dynamics of the home see LOYD (2009).

pensate for what neglected children were not getting at home. The general opinion was "a five-year-old disadvantaged child, due to environmental deprivation, was perhaps at the same level of development as a three or four-year-old middle-class child" (COONEY 1966, 17). The report looked into the uses of open-circuit television to "simulate the intellectual and cultural growth in children of preschool age". Importantly COONEY's report directly dismissed the federal Head Start program and National Education Association proposals that all children should be able to go to school at the age of four. In COONEY's estimate this proposal would only "encounter staggering obstacles", as there were not enough schools or teachers, and the costs of putting four and five years olds into the school system was estimated at 2.75 billion not even counting the building of new classrooms. Yet preschool education was urgently needed. COONEY posited "we must begin to search for new means and techniques to solve our educational problems"(COONEY 1966, 7). This new means of education would be modelled after commercial advertising, and the technological medium would be television.

COONEY's report had a huge impact. Morrisett's connections were vital in securing the financial backing of their project. For the Carnegie Foundation, the project appealed to many aspects of their mandate during the time: television, education and poverty reduction (LAGEMANN 1989). COONEY pitched the idea of preschool educational television vigorously and partnerships were made with the program to be aired on 170 stations. The audience was projected to be half of the nation's 12 million children. By February 1968 they had secured around eight million in funding; around six million from Carnegie Corporation and the rest coming from the Ford Foundation, the National Centre for Educational Research and Development in the US Office of Education, The US Office of Economic Opportunity (the major funder of the War on Poverty), the National Institute of Child Health and Human Development, Corporation for Public Broadcasting, National Foundation of Arts and Humanities, and the John & Mary Markle Foundation. Many of *Sesame Street*'s funders were the same ones who were financing cable communications research (see LIGHT 2003, 189-190). *Sesame Street* would secure additional support over the years through Morrisett's extensive networks with MITRE, the Urban Institute, the RAND Corporation and the Markle Foundation (LIGHT 2003, 204). In March 1968, the Children's Television Workshop (CTW) was launched at a press conference held by the Office of Education, Carnegie Corporation and the Ford Foundation. The CTW was framed as an experiment between television professionals and educators, psychologists, and child development specialists to "close the gap" between disadvantaged and middle class children (MORROW 2006, 65). With the experiment funded, staffed and ready to go, the show just had to be produced.

THE CONSTRUCTION OF *SESAME STREET*

The whole thing...has been constructed to bring home a sense of naturalism – an old building, several empty lots along the way getting ready for urban renewal/Black folks removal, a candy store, all that...Gordon, is a nice, solid Black male image thing that I would guess is pleasing to kids and parents who watch him daily (RILEY 1970).

E is for Eyes on the Screen

The CTW organized a series of brainstorming seminars during the summer of 1968 to articulate the goals and methods of the show. During this production process equal time was given to the content of the show and the audience's viewing experience. Both the childhood education experts and the show's producers were generally pessimistic about the average home as a learning environment, particularly disadvantaged homes. From COONEY's initial report:

> These homes are apt to be overcrowded; there are usually a large number of children in the family; the television set is on from early morning until late at night and simply one more thing contributing to the din and confusion characteristics of most impoverished homes (COONEY 1966, 45).

In these overcrowded homes the show needed to compete with the din and therefore required intense visual stimulation. COONEY hired Ed Palmer to head the internal research group of CTW because he brought to the team his "distractor", a device that quantified how, why, and when children looked away from screen. The coming together of intensive child science and television production became known as the CTW model (MORROW 2006, 68–70, 79). Jim Henson, an independent puppeteer during the summer of 1968, was brought in to keep the children's attention and establish a sense of fantasy. After some initial productions were made in June 1969, CTW paid 100 families in inner city Philadelphia one hundred dollars to watch five shows. The children's reactions were studied to quantify the number of times the child looked away from the screen. The aim was to maximize children's "eyes on the screen". This testing would ensure, through measurable results, that the show would appeal to the urban poor (MORROW 2006).

A is for Affective Skills

COONEY was especially worried about children's well-being and their emotional adjustment, how *Sesame Street* could influence the development of their "affective skills". From COONEY's first report: "Another goal which I would include is beginning awareness of basic emotions (aggression, fear, etc.) as a step toward mastering them" (COONEY 1966, 23). The Children's Television Workshop aimed to better understand the aggressive nature of children, and how children could be taught to become less aggressive through intellectual development. COONEY said looking back in fact…we have affective goals in the show, all the time, probably that is the most important contribution we've made, treating people in a kind way, in a co-operating with each other, this is what we call affective skills" (COONEY 1998, April 27).

Gerald LESSER was brought in as a research psychologist, becoming the primary academic on the project and a central figure in CTW (for his history see LESSER 1974). LESSER pushed CTW to "be direct and bold in exploring emotional problems and social issues" (POLSKY 1974, 72). The first seminar for *Sesame Street* in 1968 was called "Social, Moral, and Affective Development" and the seminar's report stated:

> There are risks in dealing with emotional material. Any issue or emotion, such as fear, is loaded…In presenting emotional subjects, the child's anxieties may be increased…It is less risky to teach about issues and emotions, not through content, but in context and through the life styles of the people on the programme…through this indirect route the programme might also address itself to supplying some of the elements usually missing on television; real fathers, Negro men and women who love one another" (in POLSKY 1974, 77; also quoted in BECK 1979, 42).

The progression of these affective skills in children was quantified for CTW by the independent Educational Testing Service (ETS) in the form of what COOK et al. called the "Emotions Test". The "Emotion Test" evaluated how children recognize and label happiness and sadness after viewing the show. These learned abilities were then used to assess children's attitudes towards race, school, and other children (COOK et al. 1975, 178). At the same time *Sesame Street*'s goal of awareness of the "Social Environment" was seen as how the child identified themselves in relation to others who had "role-defining characteristics" including such authorities as the mayor, policeman, fireman, solider; and with institutions and social settings like the family, the neighborhoods, and the school. The goal was to teach children "to see situations from more than one point of view, [to] begin to see the necessity for certain social rules, particularly those insuring justice and fair play" (BALL and BOGATZ 1970).

I is for Integration

Michael HARRINGTON's book *The Other America* suggested that mobilizing support for action against poverty required methods of overcoming ignorance about its nature and extent in U.S society. In the past, he argued the middle classes had to pass through the "Negro ghetto or the blocks of tenements, if only to get downtown to work or entertainment". Today, however,

> the very development of the American city has removed poverty from the living, emotional experience of millions upon millions of middle-class Americans. Living out in the suburbs it is easy to assume that ours is, indeed, an affluent society.

Broad support for tackling inequality within the context of not only social but spatial segregation required a means of making the poor visible to the middle classes.

Sesame Street never explicitly aimed at bringing the plight of the poor into middle class consciousness. Nevertheless – given its subject matter and social mandate – it did make a particular version of HARRINGTON's "economic under-world" visible to millions of American viewers. Because its producers felt that a show focused exclusively on the urban poor would only normalize segregation by reproducing it on the television, it choose to situate its social mandate within a racially integrated cast. While progressive in its intent, this choice meant that *Sesame Street*'s representation of inner city life was far more akin to Jane Jacob's urban fantasy land than the realities of everyday life for most of the urban poor.

Sesame Street's primary goal was to bridge racial and class divide by raising educational levels of low income preschoolers within the broader context of fighting poverty and improving the quality of children's television. While not aimed exclusively at any one social group, given the racialized nature and spatial distribution of poverty in the U.S, the "bulls-eye of the target", as COONEY put "was…inner city poor four year olds" (COONEY 1998, April 27). The show's producers however were worried about replicating stereotypes and sustaining social divisions by creating a 'segregated' television show. Moreover, given recent civil rights legislation such segregated programming appeared counterproductive and perhaps even illegal. The solution, the producers decided, was to craft a curriculum for everyone, but make an extra effort through the choice of set and characters to ensure the show appealed to the urban poor. From the outset then, *Sesame Street* would have a racially integrated cast: black actors would assume leading roles, but whites would be present and active as secondary characters. Importantly, the show typically projected an image of racial harmony: the real material divisions and social hierarchies between blacks and whites in the U.S. that had prompted the very genesis of the program remained invisible to viewers.

G is for Ghetto

COONEY was adamant that the set design must not alienate the 'disadvantaged child'. Yet, while much research was conducted on curriculum development, the producers appeared to rely on a stereotypical 'ghetto' aesthetic when recreating a realistic inner city setting. For instance, the show's original and long time creative director and executive producer Jon Stone was responsible for the set design. Stone came from a wealthy New Haven family and was educated at Williams College and Yale before joining the Children's Television Workshop. He aimed to recreate "a real inner city street" where "real people" lived. Stone's rationale for the streetscape set was "inner city disadvantaged kids are so often housebound, stuck home all alone while their mothers work, and God knows where the fathers are…They look out the window and that's where the big kids are playing and the noise is going on. The street outside is where it's really happening…where the action is". Childhood education experts and psychologists re-emphasized their recommendations that the reality aspect of the street be maintained. When the show was launched in 1969, the *New York Times* remarked:

Sesame Street could be any street in East Harlem. It has an aged brownstone, tin-corniced stores, trash cans, a vacant lot fenced in by reclaimed wood doors...broken sidewalks and a backyards crisscrossed with drooping clotheslines. Actually it exists on a sound set at Broadway and 81st Street (FERRETTI 1969).

Thus, to reach urban preschoolers the producers recreated "a typical street in a ghetto area in a major city" (WINDELER 1969). Yet while the set was aesthetically 'authentic', this streetscape was more urban fantasyland than realistic inner city. Cognizant of the need for the show to appeal to a population beyond the inner city in order for it to survive, COONEY was hesitant to represent the everyday landscape of the inner city poor, in part because no other mainstream show up to that time had been set in such a locale. Given the extreme conflict and violence most middle class Americans had come to associate with inner city, such an 'authentic' representation seemed too large a production risk. While educational experts strongly advised that the creators needed separate the humans from the Muppets in order for preschoolers to differentiate between fantasy and reality, the producers instead mixed large amounts of fantasy, play and animation on the 'real' inner city set.

R is for Role Models

All *Sesame Street* characters were based on curriculum ideas. For instance, producers thought Gordon, an African American male and one of the show's primary hosts, could serve as a positive father figure to African American children. As COONEY put it:

Ghetto kids like TV as much as other kids. Visually they're very hooked...But they never see themselves on the screen. Television's got to begin reflecting them. There are no black cartoon figures to identify with. Why not tiny Harry Belafonte type? 'Black is beautiful' is a gorgeous slogan. Why not show blacks in beautiful situations doing beautiful things? (in MORROW 2006, 66).

Gordon, the show's main character would fill the so-called masculine void that Daniel Patrick MOYNIHAN, in his hugely influential 1965 report, *The Negro Problem*, had identified as a systemic cause of the 'culture of poverty' among blacks (MOYNIHAN 1965). If Gordon was meant to project a positive image for young black children, he also represented a non-threatening image of black masculinity to which white middle class viewers could relate.[4] For instance, the very first episode of *Sesame Street* opens with Gordon showing his *neighborhood* to a white child whose family had just moved in down the street. Gordon's persona, actions and goals didn't conflict with liberal ideologies, and like white middle class suburbanites, he also aspired to the American dream. At the very least, Gordon and his wife Susan, as the show's primary hosts would – in the wake of desegregation legislation – expose white children to side of black urban America

4   For a contemporary example such a black masculine exemplar see COLLINS (2005).

that wasn't being represented in prevailing media images of the era. Thus, for instance, after an extended viewing, one teacher paraphrased a student who said: "Susan and Gordon are not really funny or bad people. They still have different skin and hair, but now I know them…" (CULHANE 1970).

Unlike Gordon's character who was deferential and friendly, Oscar the Grouch was created to dramatize tolerance to difference; to give children an early experience with coping with racial diversity in the context of school desegregation. Yet the producers couched the specifics of conflict not in explicitly racial terms, but rather in personal and bodily traits such as a matter of good taste, courtesy, and manners. Many black viewers responded in the media with outrage, seeing Oscar not only as a personified stereotypical image of blackness, but as an inner city character who passively accepted his poverty, just as his neighbors equally passively accepted his lack of a home. If the producers' intention was to change the structure of inequality in the United States, Oscar's character paradoxically suggested that some people should simply accept their 'fate'. As one black daycare director put it: "That cat who lives in the garbage can should be out demonstrating and turning over every institution, even *Sesame Street*, to get out" (cited in MORROW 2006, 152).

Another criticism of the show's characters was of its unrealistic portrayal inner city youth. When conflicts around social difference arose, as for instance in the children's interactions with Oscar, they were generally muted, their solutions couched in personal rather than systemic or structural terms. Such a portrayal was largely because the show's producers aimed to ameliorate social divides in the U.S by projecting inner city children in a positive light. They also were worried that if the show became too controversial and too closely associated with racial and class issues, its educational message would be compromised since the medium could be too easily 'switched off'. As SEDULUS, the TV critic of the *New Republic* declared: "Nobody on '*Sesame Street*' is ever miserable, terrified or exultant. Problems cure themselves. '*Sesame Street*' doesn't provoke real emotions" (SEDULUS 1970; in HAY 2003, 26). As MORROW puts it:

> CTW had envisioned a fantastic place [in *Sesame Street*], urban in appearance but divorced from the systemic problems of impoverished neighborhoods…[where] conflicts had been displaced into an interpersonal realm among the residents.

Paradoxically, many experts in children's education maintained that this unrealistic portrayal was precisely where the show failed in terms of its social, moral and affective goals. According to Urie Bronfenbrenner, co-founder of the Head Start Program and professor of child psychology at Cornell University:

> The children – whether black, white, or brown – are charming, soft-spoken, cooperative, clean, and well-behaved. When they speak, they say the right things. They never fuss, fear, or fail. Among the adults, there are no cross words, no conflicts, no difficulties nor for that matter, any obligations or visible attachments. It is to be hoped that the talented artists and researchers…will now strive to make of its second season a neighborhood – and a society – where living is human (in POLSKY 1974, 78).

CONCLUSION: SESAME STREET, JANE JACOBS, AND THE CULTURAL
PRODUCTION OF A BOURGEOIS UTOPIA

Contrary to prevailing images of the city circulating in the midst of urban
America in the 1960s, *Sesame Street* inserted a vision of urban life as safe, playful, and non-threatening, not unlike the street ballet envisioned earlier that decade
by Jane JACOBS and other leading urban intellectuals of the era (see NICOLAIDES
2007). As Marshal BERMAN pinpoints in his critique of JACOBS, 'what makes
[JACOBS'] *neighborhood* vision seem pastoral…is [that it's] the city before blacks
got there' (BERMAN 1982, 323). While African Americans occupied a central
place in *Sesame Street*, the show nevertheless inserted a pastoral vision of urban
life into homes across the U.S. This vision was squarely situated in a broader discourse of the era that pathologized the problems in cities to the black 'ghetto',
rather than to the beliefs, actions and inactions of homeowners, businesses and
governments that held the power and means to generate urban decline and crises.
As solutions to these problems came to life on the television screen, they remained far removed from the realities of everyday life for the racialized urban
poor.

Perhaps unsurprisingly then, the show was subject to a range of critiques.
These critiques mobilized existing pressure on the government from marginalized
groups to regulate the broadcasting industry to be more inclusive. Evelyn Davis,
for instance, the African American head of CTW outreach program, summarized
protests over *Sesame Street's* lack of Hispanic characters:

> [A]fter the riots…the question was 'How are we going to keep the black folks quiet?' The
> Latinos were saying, 'Well we didn't riot and we have even less. We've been totally left out
> and we are not going to be left out anymore'…All segments of the population thought they
> needed to be represented, and that their children had needs that should be addressed. And
> since the money was federal, they thought that had a right to insist upon inclusion (in DAVIS
> 2008, 216).

Feminists too demanded the producers respond to *Sesame Street's* perceived
"tokenism" and "relentless sexism" (BERGMAN 1972). While the producers did
respond in certain ways to these critiques – transforming for instance, Susan from
a stay at home mom to a working one – they did not want to compromise the
overall goals for the show's target audience by diminishing the presence of a
strong male role models for young black children, particularly boys. COONEY
countered the National Organization of Women's claims that the show's female
characters were too stereotypical, stating that "[w]e consider our primary aim of
reaching and teaching the disadvantaged child a life and death matter, for education determines whether these disadvantaged youngsters enter the economic mainstream of American life or not" (DAVIS 2008, 214). She re-asserted that *Sesame
Street* was "funded to serve the needs of disadvantaged children in this country,
with particular emphasis on inner city poor" and their needs must come first. If
they failed these children would remain poor and the Great Society would be un-
realized. Thus the primary strategy adopted *Sesame Street's* producers was what
critical media theorist term 'assimilation' (MONTGOMERY 1989). In other words,

they responded to advocacy groups demands for more inclusive and representative programming, but they did so in ways that incorporated these demands to existing social scripts, overarching narratives and a set aesthetic designed to capture the attention of their target audience while not alienating mainstream viewers. This was a general technique of placating activists developed by the television industry during the 1970s and 1980s to minimize political disruptions in the production schedules and business cycle. Television producers would 'assimilate' dissent and critique (MORROW 2006, 152).

What then did *Sesame Street* teach? In its efforts to improve the educational levels of the poor, the producers adopted a framework for change that inserted a vision of the city that would appeal to privileged viewers while attempting to placate those who had the most at stake in such a revisioning. A sanitized 'ghetto' streetscape became the site from which emotional skills were taught. Such social setting – absent of violence, depression, loss, and evictions – helped to quell middle class suburbanites fears about the ghetto. Ultimately *Sesame Street* was part of a broad cultural revisioning of urban space during the late 1960s and early 1970s into an image fit for bourgeois consumption.

REFERENCES

BALL, S. and G. A. BOGATZ. (1970): The first year of Sesame Street: an evaluation; a report to the Children's Television Workshop. Educational Testing Service.

BEAUREGARD, R. A. (2003): *Voices of decline: the postwar fate of U.S. cities,* New York.

BECK, T. K. (1979): Widening Sesame Street. *Learning, Media and Technology,* 5, 39–43.

BERGMAN, J. (1972): Are Little Girls Being Harmed By 'Sesame Street'. *New York Times.* New York.

BERMAN, M. (1982): *All that is solid melts into air: the experience of modernity,* New York.

COLLINS, P. H. (2005): *Black Sexual Politics: African Americans, Gender and the New Racism,* London.

COOK, T. D., H. APPLETON, R. F. CONNER, A. SHAFFER, G. TAMKIN and S. J. WEBER (1975): *"Sesame Street" revisited.* New York.

COONEY, J. G. (1966): The Potential Uses of Television in Preschool Education: A report to the Carnegies Corporation of New York. New York, Carnegie Corporationa of New York; Children's Television Workshop, New York.

COONEY, J. G. (1998, April 27): Archive of American Television: interview with Joan Ganz Cooney. WERSHBA, S. (Ed.): New York City, www.emmytvlegends.org/interviews/people /joan-ganz-cooney.

CULHANE, J. (1970): Report Card on Seasame Street. *New York Times.* New York.

DAVIS, M. (1990): *City of quartz: excavating the future in Los Angeles,* London.

DAVIS, M. (2008): *Street gang: The complete history of Sesame Street,* New York.

DORFMAN, A. and A. MATTELART (1975): *How to read Donald Duck: Imperialist ideology in the Disney comic,* New York.

DREIER, P. (2006): Jane Jacobs' Radical Legacy. Summer.

FERRETTI, F. (1969): 'Sesame Street' Rolls Out TV Red Carpet for ABC's. *New York Times.* New York.

FLORIDA, R. (2002): *The rise of the creative class: and how it's transforming work, leisure, community and everyday life,* New York.

GRAHAM, H. D. (1980): On Riots and Riot Commissions: Civil Disorders in the 1960s. *The Public Historian,* 2, 7–27.

HARRINGTON, M. (1963): *The other America: poverty in the United States,* Baltimore.

HAY, S. A. (2003): "Sesame Street" and the Media": The Environments. Frames, and Representations Contributing to Success. *College of Communication.* Ohio University 142, 26.

JACOBS, J. (1961): *The Death and Life of Great American Cities,* New York.

KERNER, O. (1967): *Report of the National Advisory Commission on Civil Disorders,* Berkeley, Calif., University of California.

LAGEMANN, E. C. (1989): *The politics of knowledge: The Carnegie Corporation, philanthropy, and public policy,* Middletown, Conneticut.

LESSER, G. S. (1974): *Children and Television: Lessons from Sesame Street,* New York.

LEY, D. (1996): *The new middle class and the remaking of the central city,* Oxford.

LIGHT, J. S. (2003): *From warfare to welfare: defense intellectuals and urban problems in Cold War America,* Baltimore.

LOYD, J. M. (2009): "War is not healthy for children and other living things". *Environment and Planning D: Society and Space,* 27, 403–424.

MCCONE, J. A. (1965): Violence in the city: an end or a beginning? A report by the Governor's Commission on the Los Angeles Riots, in R. M. FOGELSON (Ed.) New York.

MONTGOMERY, K. (1989): *Target: Prime Time, Advocacy Groups and the Struggle over Entertainment Television,* New York.

MORROW, R. W. (2006): *Sesame Street and the reform of children's television,* Baltimore.

MOYNIHAN, D. P. (1965): The Negro Family: The Case For National Action. Office of Policy and Planning Research, U. S. D. O. L. (Ed.).

NICOLAIDES, B. (2007): How Hell Moved from the City to the Suburbs. KRUSE, K. and T. SURGRUE (Eds.): *The New Suburban History.* Chicago.

PALEN, J. J. and B. LONDON (1984): *Gentrification, displacement, and neighborhood revitalization,* Albany.

POLSKY, R. M. (1974): *Getting to Sesame Street: origins of the Children's Television Workshop,* New York.

RILEY, C. (1970): A Black Father on 'Sesame Street' 'It's Into Cognition, Baby'. *New York Times.* New York.

SEDULUS (1970): Sesame Street. *New Republic,* 27, 23–28.

SMITH, N. (1996): *The new urban frontier: gentrification and the revanchist city,* London ; New York.

SMITH, N. and P. WILLIAMS (1986): *Gentrification of the city,* Boston.

SUGRUE, T. J. (1996): *The origins of the urban crisis: race and inequality in postwar Detroit,* Princeton, N.J.

WINDELER, R. (1969): Rich TV Program Seeks Youngest. *New York Times.* New York.

Steven M. Graves

MOVIN' ON UP: GENTRIFIED LANDSCAPES ON TELEVISION

INTRODUCTION

This work started as a failed attempt to demonstrate the operation of SMITH's (1996) "rent gap theory" in Cincinnati many years ago. Mt. Adams, Cincinnati's most fully gentrified neighborhood at the time, never appeared to have gone through a cycle of disinvestment suggested by the theory. The neighborhood known as "Over-the-Rhine" turned out to be the more likely candidate for the rent gap scenario to play out, but it began gentrifying only in the 1990s, a decade or two after the Mt. Adams area gentrified. Housing values never really faltered in Mt. Adams. The neighborhood was always pretty nice. This is not to say however that the potential ground rent did not skyrocket in Mount Adams. The difficult part of the equation is understanding why the potential ground rent for housing in Cincinnati's various neighborhood changed at different rates in different neighborhoods. Why did Mt. Adams, clearly less a bargain than the other candidates for gentrification, attract gentrifiers long before competing neighborhoods around Cincinnati?

I do not seek to seriously engage SMITH's rent gap theory here – certainly that has been done elsewhere. Rather, I want to offer an alternative perspective to the question of how the potential ground rent in SMITH's formulation came to rise so rapidly during the early 1970s. I propose that popular television programs airing during the early 1970s were a significant, silent, contributor to the successful marketing of gentrified housing to young urban professionals during this era. Several widly popular TV programs airing during the 1970s helped bolster the potential ground rent of specific types of urban housing stock by altering the symbolic value of these housing types. Demand for late 19th and early 20th century Victorians, Brownstones and the like was bolstered by TV programs that housed highly charismatic television characters, who were often young, chic, professional and childless in these types of homes. Simultaneously, television producers frequently placed TV families from other points in their life cycle or career arc in landscapes and housing stock that were clearly not gentrified. In short, television programming has been a significant factor in the generation of demand for specific types of housing, and perhaps the diminution of demand for other types of housing stock. Television made gentrified living appear highly desirable without ever exposing the many hidden costs associated with the gentrification process.

DEMAND SIDE EXPLANATIONS

A number of demand side explanations have been offered by authors who have likewise sought to explain gentrification. My work here could be considered an addition to some of the consumer sovereignty-friendly positions that have been forwarded by other authors (e.g. HAMNETT 1991; LEY 1996). There is little doubt that changing patterns in the economy, long commutes, the boredom associated with suburban living, and suburban ostracism made inner cities more attractive for young people in the housing market. However, these conditions are insufficient to explaining gentrification as it happened.

Multiple neighborhoods were available to the young urban professionals who chose not to live in the suburbs. The best of these options, from a spatial perspective, seems to have been the inner city, but even within the inner city several housing options faced potential gentrifiers in the late 1960s and early 1970s.

The most available inner city housing options for young urban professionals wishing to live near downtown depended in part on the city itself, but in general there were glass and steel, modernist high rises; older 19th century apartments and townhouses; 19th century tenements and modernist public housing or 'projects'. Other options could have been imagined, but weren't. It is clear that some of the hopes (or propaganda) that accompanied various government sponsored urban renewal programs stand testament to the fact that these other options, such as single-family detached homes, were not seriously considered. Clearly some housing options were more appealing than others in the late 1960s, but late 19th century apartments and townhouses, along with a few 19th century tenements and some early 20th century models evolved as the preferred housing stock for gentrifiers in most US cities.

TASTE PREFERENCES

A silence in the literature on gentrification surrounds the genesis of the taste preferences that led to the emergence of readily recognizable, almost formulaic, gentrified neighborhoods. How does one decide where to live, or what their house should look like? For the poorest among us, those choices are limited, but for those with better means the possibilities are more numerous. Buying a home, is generally the largest personal investment one makes *and* it is often a significant personal statement about oneself (see e.g. ADAMS 1984). Renters no doubt seek housing stock that suits their personal taste preferences as well. A home is a form of communication and the message it sends will be communicated properly if the persons for whom the message is intended can understand it. Successful transmission of the messages surrounding housing is a fundamental form of landscape literacy.

The young urbanite who sought to avoid the suburbs in the early 1970s had several urban housing options with which to make a personal statement, but only a couple of the housing options available promised any sort of positive, discernible communicative power. An apartment in a new glass and steel, high-rise building would have sent a different message than a split level home in the suburbs in 1970 or a duplex in a working class neighborhood just beyond the inner city. What housing styles say about the persons who occupy them is in constant flux and open to contest like any other cultural signifier.

Until about 1970, the housing stock that would form the common vocabulary of gentrified neighborhoods conveyed a garbled message to most Americans. Buying a dilapidated Victorian house in a neighborhood long-since abandoned by any sign of the middle class would have made a nearly incoherent personal statement the 1950s and much of the 1960s. Similarly incoherent messages would have likely been sent by middle class professionals seeking to set up a home in run-down brownstones, decrepit tenements, decaying shotgun houses, or any other housing stock that had seen better days. Clearly, the neighborhoods and housing that had been neglected and abandoned by the middle class families who once resided in them had become incapable of making a positive personal statement about the people who lived in them.

The inability to make a clear personal statement with run-down housing stock would seem to explain in part why artsy, bohemian types are often credited with pioneering gentrification in many inner city areas. People from the arts community, gays and lesbians, hippies and other folks from subcultures outside the mainstream are accustomed to making personal fashion statements that are incoherent to those in the mainstream. Music, fashion, theater, cuisine and other taste preferences are frequent sources of miscommunication between bohemians and the suburban middle classes. Often those messages are purposefully constructed to be antithetical to the mainstream. It is also completely reasonable to think that the "starving artists" who are counted among early gentrifiers, had little choice in their ability to make a personal statement; so they sought to wrest control of the message conveyed by these home and transform it into something more positive.

The question that remains though is how did so many of these colonies of bohemians and starving artists become expensive, exclusive, clearly articulated and clearly understood statements about neighborhood and home? How did consumer taste preferences for housing change such that urban residents, many of whom have been identified as "upwardly mobile" and "engaged in the "new economy," become willing to take a chance on the homes that a generation earlier had been abandoned to the under classes? How did the housing stock, which only a few years earlier communicated a message of unsavory decrepitude somehow become a recognizable signifier of a lifestyle that was trendy, urban and smart?

These questions have been explained away too easily by advocates of consumer sovereignty positions. The few times I've asked this question from those who seem well versed in the theoretical battles surrounding gentrification have dismissed this question. I've heard more than once, "Oh it was *Better Homes and Garden*, *City Living*, and those types of magazines that built the taste preferences for gentrified housing." Those sorts of explanations ring hollow. Other consumer items, including many that are far less conflated with one's *personal* identity as a home, are not be marketed so breezily. Massive marketing campaigns are launched to craft taste preferences for a wide array of items, and they often result in spectacular failure. Pop music albums, for example, exhibit powerful tendencies to sell poorly in the face of massive marketing campaigns, while other musical groups sell millions of albums despite meager marketing efforts (see e.g. DENISHOFF 1975; FRITH 1988; WICKE 1990; DANNEN 1990, 16). If the music industry has been unable to control taste preferences for something as relatively cheap as a CD, it seem very unlikely that a lifestyle complex as significant as gentrified neighborhoods could be marketed so effectively by a handful of magazines whose marketing prowess was considerably more inconsequential than that arrayed by the media conglomerates like Sony.

Since radio stands as the most, albeit partially effective means for marketing music, it stands to reason that an especially powerful media, such as television, would be necessary to market commodities that function as as the most important markers of personal or cultural identity, especially one as costly as a home.

At the very least, televisual imagery of housing structures consumer taste preferences by creating knowable categories into which home-buyers can project themselves. Television programs did not *create* gentrified housing as a knowable category; that went on for hundreds of years before television. Moreover, very influential gentrification movements began in New York and San Francisco shortly before they begin appearing on television. What television did however was to promote and regularize "gentrified housing" as a knowable and desirable category of housing, especially for young, urban professionals or those who at least aspire to be upwardly mobile, hip or fashionable.

What follows presents a brief evolutionary history of urban and suburban housing as it appeared on TV since the 1950s. A modest typology of housing options and the attendant life cycle associations is forwarded in order to make the case that TV programs have structured the evolving vocabulary of urban housing. In other words, a case is made that certain types of TV families (or characters) are placed in certain types of housing over the years; and that these categories change through time. These televised domestic settings helped viewers understand what type of homes were appropriate and desirable for what type of person. The criteria for inclusion in the analysis was only that the program was domestic, urban or suburban and ranked among the top 20 programs, according the Nielsen Ratings each year from 1950 to 1990. Using 1970 as the seminal year for American gentrification, the 1960s and 1970s were given additional consideration.

1950s

There were relatively few programs in the 1950s with regular domestic settings. Westerns, game shows and variety programming were most common and most popular. The domestic comedy had not yet fully flowered, but programs such as *Father Knows Best* and *Dennis the Menace* fueled the idea that the suburbs with single family homes on large lots were the domain of middle class, white, nuclear families. Exterior shots of the home and neighborhood, so commonly used after

1970, were rare in the 1950s. Not far removed from stage traditions, the home setting, characters and even the actors were subject to change without warning and sometimes without explanation.

Several programs during the 1950 were especially powerful templates for domestic comedies that later became staples of TV programming. *I Love Lucy*, which was extremely popular during the 1950s, frequently used domestic sets. Early seasons found the characters in an apartment on the Upper East Side of Manhattan, presumably since Lucy's husband was in the entertainment industry. Later seasons found Lucy's family living in a Connecticut suburb; which presumably was judged to be a more appropriate location for growing family. A closely related domestic sitcom was *Make Room for Daddy* (later known as *The Danny Thomas Show*), which like *I Love Lucy*, featured a performer-father who lived with his family in a Manhattan apartment in the early seasons. This family also moved to a suburban setting in later years. The other popular domestic comedy of the era, *The Honeymooners*, depicted urban, tenement-style living, making this program a rare and effective counterpoint to the more common suburban setting. This program did little to glamorize urban life. The chief characters were all childless, so they do not seem to warrant inclusion in the suburbs. The evacuation of two of the most famous TV families to the suburbs foreshadows the wholesale abandonment of the inner city by TV families in the 1960s and seems to parallel the so-called 'white flight' occurring in the post war era in the actual cities around the country.

1960s

Westerns, game show and variety programming decline somewhat in popularity throughout the 1960s, and situational comedies grow in popularity. A large number of the sitcoms of the 1960s were set in rural locations in what could be interpreted as another commentary on the decline of the American city. The *Andy Griffith Show* and *Green Acres* lead a variety of small town and rural-set TV shows during this decade. Among the domestic comedies that were set in the city, suburban housing is the rule for nuclear families. Programs such as *My Three Sons, Bewitched, Hazel, The Dick Van Dyke Show*, etc. were very popular. Even some of the programs from the 1950s, like the Lucille Ball and Danny Thomas vehicles were re-set into suburban neighborhoods. Inner city locations were featured in many TV programs during the 1960s, but almost without fail, big cities were utilized as a setting for crime dramas and detective programs. The urban set sitcom almost becomes extinct during by the mid-1960s. This reflects a growing sentiment that the city proper is a poor place to raise a family, even a TV family.

Several of the programs that seem in retrospect to be well suited for a more urban setting, were nevertheless set in the suburbs. For example, *I Dream of Jeanie* was set in a a ranch home in Cocoa Beach, Florida, no doubt a by product of the main character's occupation as astronaut, but in retrospect, it seems odd that a bachelor astronaut would live by himself in a suburban neighborhood. These unusual settings suggest that there was uncertainty among the producers of television programmers about what to do with bachelors, young couples and other would-be gentrifiers. It appears that TV producers and script writers did not yet have a viably communicable category of urban housing within which to house specific types of characters.

Only a few domestic comedies were not clearly not in the suburbs. *The Patty Duke Show* was a popular comedy that followed used the wacky-kid-has-a-problem-and-it-gets-resolved formula so very popular in the 1950s. Those scenes that took place in the home were supposed to be occurring in an upscale town-house in Brooklyn Heights. *A Family Affair* and *Julia* were the two other moderately popular, urban-set domestic programs during the 1960s. The premise of *A Family Affair*, echoing many others during this period, revolves around an unlikely family grouping led by a successful bachelor-father who is pressed into raising children in his elegant Fifth Avenue apartment. The setting for *Julia*, a rather plush suburban apartment actually generated some controversy because, according to critics, it salved the conscious of white viewers who could watch without being confronted with the realities of a single woman raising children in America's black ghettos. Julia's apartment also seemed out of step with what her character, a single-parent, nurse, could possibly afford. The similarly themed *Courtship of Eddie's Father*, was similarly set in a Los Angeles apartment. The Marlo Thomas vehicle, *That Girl* would have been a perfect opportunity to place a sympathetic, feminist character in gentrified housing. Several episodes seem to revolve around questions of housing, and she moves several times during the series, but the protagonist's housing is not used to the effect it is in later series of a similar ilk. The opening credits sequence in the early years of the series *That Girl* employed a sequence of scenes of Manhattan, much of it focused on modern, glass and steel high rises rather than the newly gentrifying neighborhoods in Greenwich Village or SoHo. Though perhaps losing an opportunity to make the Marlo Thomas character more edgy or trendy, the opening credit sequence does seem to have pioneered a scene setting technique that became commonplace in the 1970s.

By the late 1960s, the only housing arrangement for characters who would seemingly be well suited for placement in gentrified housing were either upscale apartments in either a glass-and-steel high rise or some sort of more traditional, luxury 19th century apartment. For the most part however, home settings were used to create a sense of middle class normalcy which in turn highlighted the irony of the characters' eccentric antics. In most domestic comedies and dramas, housing played the 'straight man'.

1970s

In the 1970s, housing becomes much more varied as programming becomes more thematically diverse. Housing also begins to play a far more active role crafting plot lines. Housing ceased to play the 'straight man' during the 1970s evolving into an fundamental element of many programs; creating dramatic tension, building sympathy for characters, figuring prominently in many plots and reinforcing the essential nature of TV families. It is during this period that domestic TV programs play a significant role in creating and maintaining the symbolic qualities of the gentrified home as a desirable housing option for urbanites. Television programs also helped create, alter and/or maintain the symbolic characteristics of other housing types as well, helping equate, in the minds of millions of viewers, various categories of city housing with specific stages in the life cycle, class and ethnic identity. TV tells viewers who belongs in what type of house and in which neighborhood and who does not belong.

Unlike the 1960s, where single family homes in upper middle class suburbs are predominant, the suburbs all but disappear from domestic comedies during the 1970s. A few exceptions are noteworthy and they are largely reserved for nuclear families, such as those found in *Eight is Enough* and *The Brady Bunch* (a hold over from the 1960s). *Happy Days*, one of the top rated shows during the 1970s, also used a suburban setting but because the show was set in a hyper-nostalgic America, during the 1950s, it seems to have reinforced the notion of the suburbs as a place suited only for aging empty-nesters and traditional nuclear families with children. The domestic comedy *Maude*, which drew high ratings in the early 1970s is another exception. Set in Tuckahoe, a suburb of New York City, it followed the crises of a twice-divorced, arch-liberal woman, her new husband and her live-in, adult children. The opening sequence is a series of shots progressing outward from Manhattan to the suburbs (it mistakenly used a shot of the George Washington Bridge that leads to New Jersey). This sequence, which ends with a shot of Maude opening the door to her comfortable middle class home, highlights the contrast between the mean streets of New York City and Westchester County during the mid-1970s. Maude's suburban comforts function to underscore her sometimes under-informed, feminist, social liberal views on the many controversial topics the program tackled. Her home setting also helped place Maude in direct contradistinction to her Queens-based, conservative brother-in-law Archie Bunker, from whom (in TV terms) *Maude* was spun off. Viewers in the 1970s were well aware of the "familial" relationship between Maude and Bunker.

Archie Bunker was the patriarch of the family featured in the landmark program *All in the Family*. Atop the ratings for much of the 1970s, *All in the Family* was a significant trend setter for domestic comedies and urban set programs for more than a decade. *All in the Family* chronicled the life of an aging, working class couple from Queens and their live-in, baby-boomer daughter and son-in-law. The chief protagonist, Archie Bunker, was an uneducated, bigoted, arch-conservative who was liberal only with his outspoken opinions on nearly every social and cultural issue of the era.

All in the Family was ground breaking in a number of ways, including the way housing was depicted and used to provide context. The opening credits sequence featured a sequence of shots of New York City, from Manhattan out to Queens, and down the street where the Bunker family lived. The closing shot zooms in on the modest duplex at 704 Houser Street where almost all the action in the series takes place. The use of working class housing is a significant departure from the solidly middle class homes employed throughout the 1960s for domestic comedies. Only *The Honeymooners* from the first generation of domestic comedies dared depict working-class housing with any realism. The working-class neighborhood forced Archie to confront many of the realities of big city life that his liberal sister in law Maude did not. Perhaps chief among these realities was his neighbor George Jefferson, an upwardly mobile, similarly bigoted, African-American neighbor. Archie Bunker's inability to demonstrate tolerance for almost anything slightly multi-cultural, ethnic, progressive or trendy marked him as the least likely character in all of America to live a gentrified, rainbow neighborhood. His character also helped mark working-class neighborhoods as places where aging, ignorant conservatives live, angrily defending their territory against incursions into their nostalgic vision of America.

The city and urban themes become very prominent in 1970s TV programming and *All in the Family* set the trend. The sequence of exterior shots used in the opening credits of *All in the Family* were quickly copied by essentially every domestic comedy during the 1970s, right down to the closing shot of the house or apartment. These opening sequences were used as an effective shorthand for context while simultaneously reminding viewers of the backstory of the family or character portrayed in the program. More importantly, television producers, no doubt impressed by the success of *All in the Family*, increasingly turned to inner city settings and the difficulties of urban living for story lines and series context. By 1973, seven of the top 20 programs were urban-domestic settings. In 1974, 16 of the top 20 programs were set in an urban environment. Some of those were detective-crime dramas, but seven of the top ten were sitcoms or family dramas, several of them spin offs of *All in the Family*.

Two top-rated programs of the 1970s that also chronicled the lives of working-class men in inner city environments were *Sanford and Son* and *Chico and the Man*. Both shows were set in the ethnic ghettos of Los Angeles where themes of crime and the troubles of minority communities were routine plot lines. The characters were poorer and far more sympathetic than Archie Bunker, but their neighborhoods and their housing functioned to help create sympathy for them. In both programs, the characters essentially lived in their place of work; the Sanfords in a junk store and Chico and Ed in an automotive repair shop. These programs helped paint the ghetto as not only a place for minority families, and poor folks, but also for old men who had squandered their chances for more comfortable retirement years.

Also set in a ghetto, but featuring a nuclear family was the domestic comedy *Good Times* that ran throughout most of the 1970s. The Evans family occupied much the same working-class niche as the Bunkers, but they lived in a government housing project on Chicago's South Side, presumably the notorious Robert Taylor Homes. The premise was essentially that the Evans family managed to survive an endless series of urban-specific trials with warmth and courage, punctuated by son J.J.'s antics. The desperate lack of affordable housing options was a periodic plot line in the series. The projects were depicted as dangerous and dreary, despite being populated by good-intentioned people trying to survive in the broken American dream. It depicted Fordist housing at its worst and inner city life as exciting, but cruel, especially to working class minorities.

Upscale, Fordist housing occasionally housed TV families as well, but it was almost without fail reserved for professional, middle aged, childless persons or couples. A glass and steel, modernist high rise housed Bob Newhart and his wife in downtown Chicago. A world away from the Evans family on the south side of Chicago, Bob Newhart was a decidedly uncool, middle-aged psychologist in an eponymously titled comedy that featured Newhart's wry observations about his patients. *The Odd Couple*, airing during roughly the same time period as *The Bob Newhart Show* had storylines surrounding two divorced men with wildly divergent personal habits sharing a high-rise apartment in New York City. Oscar and Felix were both middle aged professionals as well. The apartment set changes during the five seasons, but it was clearly not gentrified and would probably be best lumped in with the many of the Upper East Side arrangements of several 1960's era TV families.

Sharing many elements of other successful programs of the era was *The Jeffersons,* the most popular of the several spin-offs from Norman Lear's *All in the Family* franchise. *The Jeffersons* chronicled the antics of George Jefferson, Archie Bunker's former neighbor from Queens who had managed, through hard work and thrift, to become wealthy enough to "move on up to a deluxe apartment in the sky". The opening title sequence documents the actual move from their working-class neighborhood in Queens to the Upper East Side of Manhattan. The Jeffersons were African-American, empty nesters who lived in a modernistic, glass and steel high-rise amidst mostly white neighbors. Like the gang from *Good Times*, this cast lived in Fordist housing, but the Jefferson's home was clearly upper middle class, safe and boring, offsetting the eccentric street-wise behaviors of the George Jefferson. Their housing, like that reserved for other middle aged couples, was no place for a hipster, anyone under 30 or anyone with artistic ambitions.

GENTRIFIED HOUSING IN THE 1970s

The first program to actively use gentrified housing in a situation comedy was *The Mary Tyler Moore Show* that ran from 1970 to 1977 on CBS. It was very popular, used housing effectively, and like *All in the Family* generated several spin-offs. The protagonist, Mary Richards, was a young reporter/writer for a TV news program in Minneapolis. Though the program was largely a workplace comedy, mostly set in the newsroom of the TV station where Mary worked, many scenes were shot in Mary's apartment, one of several fitted into a 19TH century Victorian in Minneapolis.

This was probably the first show on television where the apartment became itself a source of great viewer interest. According to various fan websites, Mary's apartment was very carefully considered by the production staff before the pro-gram aired. Several episodes involved Mary's apartment; her best friend and neighbor on the show, Rhoda Morgenstern, coveted Mary's very cool bachelorette pad. Apartment envy actually keeps the two friends antagonistic in early episodes. Fans of the show are still taken by Mary's fictional apartment. The house depicted in the exterior shots of Mary's apartment remained a tourist destination in Minneapolis for decades after the final episode aired. In the last seasons of the Mary Tyler Moore Show, Mary moves out of her Victorian apartment into an up-scale glass and steel high rise, a housing option more befitting her greater maturity and elevated career status. Ironically, the exterior shot of Mary's later apartment is part of one of Minneapolis' public housing complexes. Mary Richards' trendy gentrified apartment seems to have inspired a host of similarly appointed sitcom settings and established gentrified housing as the nearly exclusive domain of young, childless TV characters who had been cast as aspiring professionals. Two of the three Mary Tyler Moore spin-offs, *Rhoda* and *Phyllis* used housing effec-tively to shape the trajectory of the storyline.

The character Rhoda moved back to New York City, living for a short time with her parents in the Bronx and later with her sister Brenda on 84th street in gentrified housing. The sisters were young, childless, struggling professionals, coping with insecurities associated with youth. Their apartment was not upscale, but often featured in exterior set-up shots (a practice that was becoming extremely commonplace among domestic comedies in the 1970s) and almost always featured some comical banter with 'Carlton Your Doorman'. Rhoda gets married in an episode that ranked among the highest rated programs of the 1970s. Once married, Rhoda moved in with her new husband Joe in a glass and steel high-rise for a few episodes, but Carlton and Brenda proved so critical to the show's success that Rhoda was soon moved into a plush, gentrified penthouse atop her sister's building on 84th street. Now married and with a step-son, Rhoda's family becomes perhaps the only quasi-nuclear family to live in gentrified housing during the 1970s; but it is short lived. Rhoda soon divorces and resumes her life as a struggling artistic type, running a window dressing shop through the next couple seasons, putting her gentrified housing once again in line with her career status and position in the life cycle.

Phyllis, the other Mary Tyler Moore spin off follows the adventures of Mary Richard's former Minneapolis landlord after she moved to San Francisco with her teenage daughter. Phyllis moved in with her mother in law in a somewhat ex-travagant upper middle class home overlooking the San Francisco Bay. The premise reprieves the classic multi-generational, non-traditional family comedy routines and reinforces the detached single family home as the proper housing for middle aged people with children.

A relatively short-lived, but popular workplace comedy that also featured regular scenes in a decidedly gentrified apartment was *Welcome Back Kotter*. The Kotter character taught a remedial class in an inner city high school to a motley group of minority students. Kotter himself had been a remedial student in the school as a teenager and had made it 'out of the ghetto', only to return as a gentrifier. The Kotter's apartment was small, but very well appointed with tasteful antiques, Victorian touches and wide stripes painted on the walls in earthy tones. The Kotter's apartment, though it was in Brooklyn, represents one of the more quintessentially gentrified TV homes. Almost cliché in its décor, the apartment reinforced growing notions of proper housing for childless, educated, 20-some-things during the 1970s.

Airing at nearly the same time was *Mork and Mindy*, another program that re-volved around the lives of a young 'couple' (the character Mork was from outer space) who lived in a restored Victorian in Boulder, Colorado. The apartment setting in this program was also very nice, had an earthy vibe and was somewhat reminiscent of the Kotter's apartment in terms of interior décor. The exterior of the house, though brick, echoed the Mary Richard's house.

The last of the popular 1970s programs to feature what is ostensibly gentrified housing was *One Day at a Time*. The apartment in this program had several bed-rooms and was constantly visited by the building super, suggesting that it was old and probably in poor repair. Exterior shots were spare after the first season, but initially followed the standard sequence of shots; but this time from the suburbs to the inner city where the family moved after the family matriarch divorced her husband. The apartment building was a multi-story brick with Italianate arched windows. Oddly enough, the interior windows were not arched. The interior deco-ration was somewhat in line with other gentrified apartments of the time, but not as chic. The life cycle of the family was a bit out of line with others, but because the women on the program were all youthful and single; it apparently made sense to producers to place them in gentrified housing.

The final top-rated program in the domestic comedy genre was *Three's Company*, airing for nearly a decade beginning in 1977. Because the lead charac-ters were young, one would think they would have been best housed in something more obviously gentrified, but the show's premise may have undermined that housing option. Jack Tripper, because of his ineptitude, had been unable to secure anything but a room at the YMCA until he moved in with two attractive ladies in an apartment in Santa Monica. Because the elderly man who owned and managed the apartment complex objected to co-ed living arrangements in his building, the threesome had to pretend that Jack Tripper was gay for the run of the series. The age and conservative nature of the building's owner and the Santa Monica setting undermined the likelihood that *Three's Company* could be set in gentrified housing.

Though it was not aired during prime time and falls outside the boundaries of the study's methodology, the children's educational program, *Sesame Street* de-serves mention for its significant contribution to our notions of gentrified neigh-borhoods. Developed in the late 1960s, *Sesame Street* was a radical departure from other children's programs like, *Mr. Roger's Neighborhood* in terms of its depiction of urban life. About one-third of the segments in a typical *Sesame Street* episode were live action scenes shot in front of what appeared to be a New York City brownstone. *Sesame Street* also had obvious signs of urban decline as well, with some quasi run-down buildings, metal trash cans (Oscar the Grouch's home) and occasional graffiti tags. Still it was always a friendly place. Perhaps more than any other TV program, *Sesame Street* advanced a vision of an ideal gentrified neighborhood; the bohemian, rainbow neighborhood filled with struggling artists, small shop keepers, loosely gendered adults and even a neighborhood grouch, which gave it a realistic patina. So socially progressive was this vision of urban life that a state agency in Mississippi considered banning the show, presumably because of its integrated cast and progressive vision of urban life.

1980s

By the time Ronald Reagan had become president, gentrification had become an established fact in many American cities and the debates surrounding its costs and benefits was in full flower. In Hollywood, small screen representations of gentrified housing had likewise become solidly entrenched as a common setting used to precipitate audience sympathy for characters who were supposed to be hip, attractive and youthful.

The most beloved TV family of the 1980s was the Huxtables who appeared on *The Cosby Show*. They could not have been more different from the Bunkers of the 1970s. They were upper middle-class, African-American, urbane, well-educated and very stylish. The patriarch was Cliff Huxtable, an obstetrician married to Clair, an attorney and they, along with several very well dressed children, lived in a nicely appointed very spacious, Brownstown in Brooklyn Heights. The show was mostly apolitical and family oriented. The Huxtables were 'America's Family' and Bill Cosby became 'America's Dad'. Their house, shown in numerous exterior shots, though actually in Greenwich Village, presented a highly desirable vision of urban living which stood in stark contrast to most urban set domestic comedies of the 1970s.

Many domestic comedies during the 1980s were set in gentrified housing, and like *The Cosby Show*, the characters did not have to be young, childless professionals. Everybody that the program's producers wanted to cast in a favorable light was housed in gentrified home. Programs like, *Too Close for Comfort*, *Kate and Allie*, *Goodnight Beantown* and *Cheers* all used gentrified landscapes, despite the fact that none of the characters really fit the classic profile for 1970s gentrifiers. Gentrified housing had become largely unremarkable.

Two very notable 1980s TV families did not live in gentrified housing. One of these TV families seemingly well-suited for gentrified housing, the Keatons from the program *Family Ties*, were actually placed in a pedestrian home in suburban Columbus, Ohio, a state that is very popular among producers hoping to stress the ordinary-ness of the setting. The Keaton parents would have been classic gentrifiers because they were hip, post-hippie, liberals, but their children were Reagan conservatives. The other prominent 1980s TV family was featured in the program *Rosanne*, which chronicled, with some realism, the humorous trials of a working class family. *Rosanne* was set in a working class neighborhood in a fictional town somewhere near Chicago, though the setup shots of the house and neighborhood were from Evansville, Indiana.

CONCLUSIONS

Television programming has played a significant role in creating an illusion of gentrified neighborhoods as an urban ideal, unburdened by the hardships of displacement. Television programs helped establish the categories of knowledge regarding common housing choices. Television assisted in characterizing concrete, glass and steel, Fordist housing as suited for those over 40 and/or the urban poor. Suburban neighborhoods, though nearly universal for all middle class and working people, regardless of life cycle or family status in the 1960s became a no man's land for TV families that were not white and nuclear during the 1970s and 1980s. Gentrified housing became a popular, desirable and knowable housing option for young, trendy urban based characters during the 1970s, thanks to programs such as *The Mary Tyler Moore Show*, *Mork and Mindy*, and *Welcome Back Kotter*. These programs helped fuel demand for gentrified housing by giving gentrifiers and developers a knowable, simplistic and desirable housing/lifestyle category.

If we can assume that televisions' enormous cultural influence did indeed affect housing markets in the United States during the period under review, then we may safely deduce a bit of TV's possible role in the operation of SMITH's (1996) rent gap theory. As TV programs built demand for the categories of housing it represented so favorably, it helped raise potential ground rent for housing stock that resembled the homes occupied by America's favorite characters on the small screen. It may have diminished the value of housing for homes that resembled less sympathetic characters. In doing so, TV would have helped excite demand for neighborhoods that looked like they could already be on TV, like Cincinnati's Mt. Adams. At the same time, developers and other potential gentrifiers would have had little trouble recognizing the yet realized potential of other neighborhoods, like Over-the-Rhine, because they contained TV-sanctioned architectural features. These observations do not fundamentally challenge SMITH's well-debated theory, but they do hopefully shed light on the somewhat hidden market forces that animate both housing markets and consumers.

REFERENCES

ADAMS, J. (1984): The Meaning of Housing in America. *Annals of the Association of American Geogaphers* 74 (4), 515–526.
DANNEN, F. (1990): *Hit Men*. New York.
DENISOFF, S. (1975): *Solid Gold: The Popular Record Industry*. New Brunswick, NJ.

FRITH, S. (1988): *Music for Pleasure: Essays in the Sociology of Pop*. New York.

HAMNETT, C. (1991): The Blind Men and the Elephant: The Explanation of Gentrification. *Transactions of the Institute of British Geographers* 16 (2), 173–189.

LEY, D. (1996): *The New Middle Class and the Remaking of the Central City*. Oxford.

SMITH, N. (1996): *The New Urban Frontier: Gentrification and the Revanchist*. London.

WICKE, P. (1990): *Rock Music: Culture, Aesthetics and Sociology*. New York.

Brenda Kayzar

INTERPRETING PHILANTHROPIC INTERVENTIONS: MEDIA REPRESENTATIONS OF A COMMUNITY SAVIOR

> Launched by philanthropist Sol Price...
>
> the City Heights redevelopment project has become
>
> a working laboratory for reversing inner-city blight.
>
> (Emmet PIERCE, Staff Writer, San Diego Union-Tribune, 2002, I–4)

INTRODUCTION

In the late nineteenth-century, ad hoc speculative development produced the early working class suburb of City Heights near downtown San Diego. The aesthetic of this first ring suburb was irreparably damaged by up-zoning, municipal neglect, and plans for a freeway extension in the mid-twentieth century. City Heights has continued to serve the city's working class populations albeit of successively lower incomes and differing life trajectories. Today, juxtaposed within this high rental-tenured and densely built landscape are notable signs of new capital investment. Media accounts relate renewed interest in City Heights to projects headed by a local philanthropic organization. In addition to a retail mall, Price Charities[1] is associated with the development of numerous other projects including a community center, an office complex, affordable townhomes, senior housing, and an elementary school.

> None of it would have happened without Sol Price's money ("an admirer" quoted in San Diego Housing Commission online article, MORRIS 2001)
>
> Some observers question whether the partnership that transformed the neighborhood can be repeated without the financial support of philanthropists like Price (San Diego Union-Tribune, PIERCE 2002, I–1).

I suggest the focus by the media on the "capital and caring"[2] of Sol Price and his charitable organization obfuscates the work of numerous heterogeneous actors

---

1   Price Charities refers to the philanthropic entities formed by Sol and Helen Price. The Price's charitable organizations include; Ben Weingart Foundation (1980), the Aaron Price Fellows Program (1991), the Price Family Charitable Fund (1995), and San Diego Revitalization Corporation (2000), and they are managed internally by a division of Sol Price's corporation (CAHILL 2006; Price 2009).

2   RENO 2003

that laid the groundwork for, or interacted with the philanthropic organization to produce many of the changes seen and experienced in the community's landscape today. The media promotes a teleological story of progress but limits a clear understanding of the actors involved and of the community itself. Therefore an objective evaluation of the outcomes and their impact on community members is difficult to discern. In contrast to media reports included in local newspapers, magazines, and a documentary produced for public television, I utilize a network framework for comprehending the roles of numerous actors; grounding my re-interpretation of the process in a decentered representation of neighborhood change. This study seeks to move beyond the discourses sanctioned by the media to offer urban scholars and planners an alternative interpretation from which to consider differential power relations and successful and problematic outcomes.

PLANNING, REPRESENTATION, AND NETWORKS

This study focuses on representations of revitalization in the media. Urban scholars often struggle to comprehend and discuss physical change in places that are always becoming[3]. Yet they are aware of how discourses surrounding pro-cesses of change can influence how municipalities understand success or failure of specific practices and outcomes, and thereby, how they appropriate and standardize those practices and outcomes within entrepreneurial efforts, plans for growth, community and general plans, and municipal codes (ABBOTT 1993). Since the practice of planning is intimately influenced by discourse, with an eye to making policy decisions, it becomes an imperative to understand which discourses gain attention and which discourses are filtered or disregarded (HILLIER 1997). In this chapter I demonstrate how media standing, when granted to a limited set of actors, shapes what becomes a commonly accepted discourse of neighborhood change; limiting consideration of the roles of other actors, their motivations, values, perceptions, and ideals of and for place.

I return, nearly two decades later, to what DUNCAN and LEY termed a post-modern approach to representation (1993). I want to expand upon the idea of de-centering, or an effort to shift "the privileged sites from which representations emanate" (7). Publishers are generally associated with the political and economic elite of the city, and in this sense, the media represent a privileged site. The dis-courses of persons given media standing, or the opportunity to speak from this privileged site, often gain attention. Other discourses are filtered (limited in their expression) or disregarded altogether. ROTH and HAAR (2006) note how mass media shapes perceptions of urban development. Therefore, the discourses dis-seminated in the media have the potential to become dominant discourses and shape future policy decisions. The dominant discourse represented in the media regarding change in City Heights suggests the philanthropic organization has

---

3   As PRED (1984) notes, places are really about change and processes.

acted as, and is perceived as a savior and that the outcomes are not replicable without Price Charities.

For this case study I utilize a network approach to introduce a multiplicity of voices; in an effort to redistribute "representational authority into multiple sites" (DUNCAN and LEY 1993, 8), and mitigate the media standing given Price Charities. There has been a lot of interest in the application of network theory across many disciplines including geographic studies, though the theory's application in research focused on localized urban phenomena is limited (e. g. see HILLIER 1997; BICKERSTAFF and AGYMAN 2009). Yet a network approach exemplifies an appropriate framework for tracing and comprehending the complex associations inherent in neighborhood change. I draw from actor network theory (ANT) to demonstrate how achievements occur as a result of relations between collectives of people and things, flowing through and within a collective of networks (LATOUR 2005). Conceptualizing processes of neighborhood change as effects of interaction and negotiation between a multiplicity of heterogeneous human and non-human actors, amid changing economic, political, social, and planning contexts illuminates the roles of the many rather than the few and enables a richer discourse about change than the one often provided in media accounts.

ANT does not envision the network building process as a linear teleological story of progress. Instead, interactions are chaotic, with actors enrolling wholly, partially, or not at all (CALLON 1991; LAW 1992). Moreover, each actor is their own network and they relate in multiple networks of activity where they are shaped and reshaped by continual interactions. It follows that networks experience fluctuations between periods of stability and instability, as actors seek to mobilize efforts. Chaos and instability in changing communities are a reality of daily life. They are reflected in incompleteness, in the effects of diffuse privatized property rights and shifting economic markets, and they contradict the teleological and linear progression offered in media accounts of neighborhood change.

Further, the network approach defines a skein of interacting relational networks that is not embedded in place (MURDOCH 1997; SMITH 2003). Instead, networks are "stitched together across divisions and distinctions" such as macro and micro, global and local (MURDOCH 1997, 322). Divisions, based in dualistic concepts of scale, distance, space and time, are dissolved through heterogeneous associations between humans and material resources. Non-human entities are initially inscribed with the intent of the actor creating them, but are then translated with the intent of the actor reading them. The concept of inscription and translation means inorganic elements such as planning documents, seminal texts, buildings, catalyst projects, and media accounts of revitalization can be considered actors in local and more importantly, non-local networks of activity. Non-human actors have the ability to shape other actors through interaction within non-proximate and time differentiated networks, stitching together national, regional and local contexts, and media accounts, as non-human actors, can gain greater influence by extending a singular discourse through the distance of time and space.

Considering the importance of the debates surrounding revitalization of older inner city landscapes, it is important for urban scholars, planners, and civic leaders to hear a polyphony of voices from neighborhoods undergoing change. Proponents laud increased revenues and greater income-diversity (GROGAN and PROSCIO 2000; CLARK et al. 2002; RYPKEMA 2003), but opponents note the exclusionary outcomes of commodification and militarization in revitalized older communities (DAVIS 1992; ZUKIN 1995) and the displacement of lower-income groups in conjunction with ongoing redevelopment and increasing levels of investment (SMITH and WILLIAMS 1986; LEES, SLATER and WYLY 2007).

> The Price family and its Price Charities began investing in City Heights when the crime-ridden area was considered such a civic problem that the City Council issued a declaration of emergency. Price's non-profit San Diego Revitalization Corps has helped raise the standard of living and lower the crime rate (In San Diego Magazine, RENO 2003).

> It's no longer a problem of people moving out of City Heights – it's about controlling the numbers of people coming in (San Diego National Bank Board Chairman Murray Galinson, quoted in Blueprint Magazine, HERMAN 2002).

In media accounts about City Heights, Sol Price states that his holistic approach to redevelopment in the community was not for the sake of development alone (MORGAN 2001; ALLEN 2004; ESPINOSA 2006). Although 'growth' tenants are an integral part of the overall plan, Price points to aspects aimed at meeting the needs and wants of the community, with an effort to empower the community. An objective analysis of the outcomes is not presented in local newspaper accounts or in the documentary, however, so the reader/viewer is left to wonder what the needs and wants of the community are, and if members in the community feel these needs and wants are being met. After an introduction to the community, I direct my focus to introducing the multiplicity of actors involved in processes of improvement in City Heights with a goal of illuminating needs, wants, and outcomes.

A HISTORY OF CHANGE AND NEGLECT

Much of the city's affordable housing advocacy has been targeted to the community of City Heights, where roughly 22% of the city's affordable units are located (SDHC 2009). Today, impacts of the national housing market downturn are evident in this older, inner-city community as well. Repossessed homes and short-sales comprise a large portion of current sales and the price per square foot for a single family home reflects an almost 50% decline in value[4] from market highs in the mid 2000s (BENNETT 2008; TOSCANO 2009).

> Price, 86, a retired lawyer and the co-founder with his son, Robert, of the renowned Price Club wholesale chain, is widely credited with the neighborhood's comeback (In Blueprint Magazine, HERMAN 2002).

---

4   By mid-2008 the price-pre-square foot had dropped to $220 (BENNETT 2008).

He's one of hundreds in City Heights who've seen their hopes for turning around their neighborhood backed by the biggest guns in San Diego's philanthropic community (In San Diego Magazine, CLARK 2002).

We need clones of Sol Price (San Diego Planning Director Gail Goldberg quoted in the San Diego Union-Tribune, PIERCE 2002, I–1).

Price Charities – the same foundation that already has revitalized large portions of City Heights (In the San Diego Union-Tribune, ENSOR 2003, M–7).

At the height of the market, media accounts often credited Price Charities for the renewed desirability by higher-income groups for properties in City Heights. The accounts highlighted the planned holistic approach to neighborhood improvement adopted by Price Charities when it enrolled in efforts to revitalize the community in 1994 (CAHILL 2006). The material outcomes of this effort are listed in Table 1, reflecting close to $400 million in investment and a mix of resources for the community from new public services, to retail and commercial spaces.

Table 1: Projects completed in City Heights after enrollment of Price Charities.

Material Outcomes: Project Type	Cost
Public Library	$14.8 Million*
Public Park	*
Community Pool	*
Community Theater	*
Community Tennis Courts	*
Community Athletic Fields	*
Rosa Parks Elementary School	$24.4 Million
4 Additional Elementary Schools (2 completed)	$200 Million
San Diego Community College District Continuing Education Facility	$8.6 Million
San Diego State University Community Center (education and job training)	$2.9 Million
Mid-City police substation	$12.7 Million**
Community Gymnasium	**
City Heights Retail Village	$20 Million
City Heights Square	Under Construction
Alternative Fuel Transportation and Education Center	$12.5
Six-story office building	$45.5 Million***
Townhomes #1	***
Career Center and Transit Plaza	$42 Million****
Townhomes #2	****
Senior Housing	Under Construction
Complied from various San Diego Union-Tribune articles and the Price Charities website.	

The investments were funded through a series of partnerships between local governance and various non-profit organizations, including Price Charities[5]. Further non-infrastructure investments by Price Charities support various educational and social service programs in the community.

> We've turned the corner where the risk capital that was put up about 10 years ago has proved successful, and private investment that wasn't willing to take the risk before is starting to come back into the neighborhood (Community Development Director for Price Charities quoted in San Diego Business Journal, ALLEN 2004, 6).

Prior to the recent nationwide recession, the infrastructure investments in the community appeared to assure potential investors that the risks once associated with City Heights were no longer insurmountable. In more recent media reports, this reduced-risk characterization of the community remains in-tact, with real estate agents implying the community will once again be a desirable place to live and work once the market stabilizes (BENNETT 2008; TOSCANO 2009). The 'positive influence' discourse of Price Charities also remains intact. In a recent newspaper article, Price Charities is noted as one of the "most active private organizations that have invested heavily in the neighborhood" (BENNETT 2008). The positive spin by local real estate agents overlays the reality of very real declines in property values and the blight of foreclosures, however, and it belies a history of problematic decision making on the part of the city with regard to the community.

City Heights is one of San Diego's first inner-city suburbs[6] (HENSLEY 1956). Comprised of modest houses on small lots the community was envisioned by early speculators as welcoming to blue collar workers, offering a privately owned street car connection to downtown. The streetcar was quickly dismantled by its operators during the real estate market crash of 1880 though, and a gap of over two decades existed before City Heights was reconnected with a new northern line (HENSLEY 1956; HEINE 1973). Due to the transit disconnect, the instability of the regional real estate market, and a topography split by finger canyons, the community developed in an ad-hoc fashion. An absence of sidewalks in many areas, in-situ development by lower-income homeowners, and the lack of contiguous streets as a result of the natural landscape gave City Heights a rural character[7], despite the proximity to downtown.

Densities in this roughly 3,000 acre area of the city increased in the 1940s. Many of the small lot single-family homes were modified with accessory dwelling units in an effort to accommodate, through the private market[8], a war-time boom in population. The increased lot densities became institutionalized through zoning amendments in the 1950s and 1960s (STEPNER 2007). Up-zoning in City Heights

---

5 Price Charities' contribution to redevelopment projects in City Heights, including for-profit ventures, was never envisioned to exceed 1/5 of total investment (CAHILL 2006).

6 The center of the community is about 6 miles from downtown San Diego.

7 What Dr. Larry FORD at San Diego State University terms a 'rurb'.

8 Civic leaders chose not to establish a local federal housing division to accommodate the production of public housing units; choosing instead to let the private market and the military provide housing for the war-time increases in population (KILLROY 1993).

reflected an effort by city planners to limit development and maintain low-densities in the periphery (CofSD 1979). Lacking city-wide design guidelines, however, the city's efforts resulted in development of poorly considered structures of inexpensive materials that were inserted randomly on lots originally envisioned for 800 to 1,000 square foot single-family homes (COLLUM 2003; STEPNER 2007).

The bulky unimaginative 'six to ten packs' fronted by surface parking spaces that comprise 29% of City Heights' housing stock[9] were aesthetically damaging to future property values (Figure 1). This aesthetic disamenity, combined with the community's aging single family housing stock and growing lower-income and minority population, subjected City Heights to the disfavor of local lenders who actively practiced redlining (FORD and GRIFFIN 1979; TUREGANO 1985). Negative impacts to the housing stock were eventually supplemented by a decline in retailing along the community's two main automobile strips. Retailers drew consumers and vitality to the area, but new suburban malls extracted both retailers and consumers from the community (SHOWLEY 1999; BUNNELL 2002).

Figure 1: The city up-zoned lots originally designed for smaller single-family homes in order to accommodate 6-10 unit apartment complexes. Photograph by author.

Finally, freeway development that facilitated access to the new suburban malls created new barriers to and within the community. The expansion of north-south and east-west by-pass routes divided or isolated neighborhoods and permanently disconnected City Heights from the grid of downtown. The most destructive freeway project was a 'paper' freeway that was incorporated into the city's general plan in 1959 (WEISBERG 1991). Construction would not start for several decades, but the presence of the 'paper' route created a path of destruction; encouraging property owners to neglect and abandon properties as they awaited the inevitability of eminent domain (FORD 2005).

As political and economic decisions reshaped the material landscape, demographic shifts reshaped the population of City Heights. The nationwide increase in foreign immigration to the United States was acutely reflected in the community of City Heights. In San Diego's hyper-segregated residential landscape, housing stock and cost prevent lower income groups such as immigrants from settling in

---

9   Census 2000

many communities[10]. Highly dense and relatively affordable City Heights accepted these populations, and the community's numbers doubled in the decades between 1970 and 1990 (MATTINGLY 2001). Poverty rates in the community also rose with almost a third of the population living below the poverty line[11]. Rental tenures rose as well in the same time period to 79-percent of households as white property owners left the area but were unable to sell their homes (Census 1970, 1990).

An often media-cited statistic suggests that over 30 languages are spoken in the community (YOUNG 1994b; CLARK 2001; STETZ 2004; GREEN 2005; ESPINOSA 2006; BENNETT 2008). According to census data, less than 1/3 of the population are 'English-only' speakers. An equally often cited statistic suggests that crime rates in the community are well above average. According to a report prepared for the Price Foundation, crime rates in 1995 were among the highest in the city (2002). There were 1,230 violent crimes reported that year, and 2,890 property crimes. The report was prepared ten years after the local store-front police sub-station was closed due to citywide budget constraints (GRAHAM 2005). The closest police station was about ten miles away. The report was also prepared a full five years after the City Council declared a 'state of emergency' in City Heights due to concerns about crime rates. The report reflects a lack of institutional initiative to bolster civic resources.

Enrolling Philanthropic Interest

A lack of public and retail services was evident throughout the community in the early 1990s. In 1993 the only major grocery store in the community closed claiming sales did not support the cost of security and insurance (CAHILL 2006; ESPINOSA 2006). At this time, media accounts note the enrollment Sol Price into City Heights networks as the capital behind a for-profit venture (YOUNG 1994c; ESPINOSA 2006). Citylink is a real estate development and investment corporation headed by former city council member, William Jones. Jones had recently achieved an MBA at Harvard and he hoped to bring retail to inner-city neighborhoods in San Diego (CAHILL 2006, ESPINOSA 2006). A commercial real estate advertisement for the shuttered grocery store attracted Jones to City Heights

---

10 A long history of exclusion through various practices from redlining to zoning has kept minorities and lower-income groups from relocating to other areas of the city (FORD and GRIFFIN 1979).

11 According the regional governance agency SANDAG, City Heights represents the greatest ethnic diversity in the city of San Diego (2003). The 2000 census showed a population of close to 79,000 where 31% self-identify as 'white', 30% claim 'some other race' as an identity, 17% claim 'Asian', 14% state they are 'black or African American', and 53% indicate they were 'Hispanic'. More recent data from the San Diego Unified School District suggest a high-poverty rate as well. Enrollment records[11] show that 100-percent of students in 10 of 12 local schools are eligible for district provided lunches (SDUSD 2009).

where he found he was in competition for the site with the city (CAHILL 2004; ESPINOSA 2006).

> The law-enforcement facility would be the first phase of a larger plan pushed by Citilink, a San Diego-based investment corporation focused on urban renewal. The corporation is headed by William D. Jones, a former City Council member, with start-up capital from Sol Price, founder of The Price Club (In San Diego Union-Tribune, YOUNG 1994c, B–1).

The city envisioned opening a police station at the site but they were unable to fund rehabilitation of the building. During negotiations regarding the site, Price and Jones created a master plan for City Heights, employing knowledge from national studies and conversations with local citizens. The proposed Urban Village encompassed a four block area and included a police station and a neighborhood retail center. Price loaned the city $3 million for the rehabilitation of the grocery property into a police station in exchange for support of the master plan, which would enable Jones to develop the retail center. The new police station opened in early 1996 at the center of City Heights. The retail center opened in 2001.

TRACING THE NETWORKS OF ACTIVITY

In most media accounts, Price's entry into City Heights is represented as an altruistic affair-as giving back to the community he lived in when he first moved to San Diego as a child[12]. Juxtaposed in the media with representations of philanthropic altruism are images of a community out-of-control. Many reports are prefaced with references to gangs, drugs, and crime[13]. In the documentary, produced as part of a series about California for public television, the set-up for Price's entry into the community is a shadowy reenactment of a police raid in an obviously abandoned home, and resident and police voiceovers describe sirens, gunshots, and police helicopters (ESPINOSA 2006).

Yet Price's entry into the community was through his support of a profitable retail venture. As the owner of the region's first big-box discount chain, Price had decades of retail experience. He also had national data to support a belief in inner-city retail. The Initiative for a Competitive Inner City (ICIC) [14] was formed in 1994 by Harvard's business school; Jones' recent alma mater, and the think tank strongly advocated for retail investment in communities like City Heights. Moreover, Crime numbers were high when compared to other communities,

---

12 For example, see ALLEN (2004); CD3 (2003); CLARK (2001); CLARK (2002); ESPINOSA (2006); GEMBROWSKI (2001); GREEN (2005); MORGAN (2001); PIERCE (2002); RICHMOND (1991); STETZ (2004); Young (1994). One journalist (BIBERMAN 2001) mistakenly suggests Jones also grew up in City Heights when in fact, he is from a community in Southeast San Diego (CAHILL 2006).

13 For example see CLARK (2002); HERMAN (2002); CLARK (2001); YOUNG (1994).

14 ICIC suggested the need for a shift in economic development strategies with regard to the inner-city (COYLE 2007).

though not necessarily when considered on a per capita basis, and the representations fail to link crime to lack of a police presence in the area, or the decaying landscape that is a result of poor planning decisions.

Media reports also consistently note ethnic diversity within the community. It is unclear how this is to be interpreted by the reader/viewer. Does this diversity somehow explain problems in the community? Does it imply that this was a community in need of a savior? MATTINGLY (2001) notes, diversity can be interpreted as both a benefit and a disbenefit. Cultural difference, the least of which is language, makes it difficult to engage across neighborhoods. Cultural difference is further complicated by instability since much of the diversity is embodied by recent immigrants to the United States living at or below the poverty level and these households move often due to economic hardships (FREEMAN 2005). In contrast, the local media do not address the lack of diversity in the rest of the city and its suburbs, or the social and structural elements of city practice that concentrate diversity into small areas of the city.

In the Price era, exclusionary practices resulting in the concentration of diversity and poverty in City Heights is not explored. Media representations merely imply an impoverished and ethnically diverse population living amid crime and degradation, ripe to receive the benefits of an intervention from an altruistic philanthropic organization. That the community was in need of public services and investment is beyond question. I argue, however, that this was not a population awaiting a savior. Instead, it was a population enrolled in networks of activity focused on improving the material, economic, social, and political contexts of their daily lives in City Heights. Diversity, actions of the state, and economic impacts simply meant that building grassroots efforts to solve problems in City Heights took time and effort[15]; in part because of limitations to community cohesiveness, but also due to the city's limited effectiveness in improving upon its limited stock of affordable housing and social services throughout the city, and inequities in the distribution of basic public services (ROYAL 2005; JOHNSON 2006; STEPNER 2007).

Tracing the networks of activity, through interviews and investigations into the pre-Price era press, reveal a number of successful efforts by various actors that preceded the involvement of Price Charities. I highlight two specific networks; the Vision Project and the City Heights Community Development Corporation, and note the importance of their influence in shaping later outcomes.

---

15 A local Assembly-person noted it is difficult to get San Diegan's involved, overall, and therefore City Heights' residents should not be stigmatized as uniquely disengaged (CLARK 2002).

Overcoming a 'Paper' Freeway

A freeway, extending nine blocks and cutting through three communities was incorporated into the city of San Diego's general plan in 1959 (KUCHER 1991; RICHMOND 1991; WEISBERG 1991). State funding was not made available for the extension of I15 until 1991 and construction did not start until the following year. City Heights was dramatically impacted by the existence of this paper street. Once construction commenced, CalTrans demolished 251 homes in the community and the loss of affordable housing was devastating, but much of the 5-block area had become uninhabitable during the 33-year wait (FORD 2005; ESPINOSA 2006). Property owners stopped maintaining structures pending the inevitable 'taking' and some properties were abandoned, reverting to city ownership due to unpaid property taxes[16]. Crime increased with vacant houses available for illicit activity, and the area become a dumping ground for discarded junk from all areas of the city.

Despite the anticipated 5-year duration of the freeway project, the community actually welcomed the start of construction (JAHN 1992; POWELL 2006). With support and guidance from the City Heights Community Development Corporation (CHCDC) they had been active in developing a plan known as the Vision Project to mitigate the impact of the freeway once construction was complete. Drawing from cities such as Seattle and Phoenix, community members wanted a freeway lid that would span a five block area. The master plan called for parks, a post office, a public library, transit stations, two town centers, housing, retail, and restaurants to be developed on the lid. This inventory represented public services and private investment badly needed by the community. The Vision Project was supported by the city council and CalTrans, and local representatives lobbied congress for federal funding (RICHMOND 1991).

A state budget shortfall of $23 million reduced the freeway lid to one city block (Figure 2) (JAHN 1992). Despite the outcome, an analysis of network interactions demonstrates that the community was able to enroll support for the Vision Plan from city and local congressional representatives. The success of freeway lids in other cities reduced CalTran's perceived risk. The state transportation agency was enrolled in planning networks where precedent for such a project had been set.

---

16 The city failed to maintain the properties under their ownership in City Heights and were called to task for their oversight (CANTLUPE 1988).

*Figure 2: A public park in City Heights developed on a freeway lid
as part of the community driven Vision Plan. Photograph by author.*

At the local level, City Heights was acknowledged as a neighborhood in 'need'. The city's redevelopment agency provided special assistance with economic development that included park development on the freeway lid[17]. Moreover, the adoption of smart growth planning principles in the city's general plan in 1979, meant civic leaders and planning staff with the city were well versed in the language of walkable town centers and transit oriented development, all of which are represented in the Vision Project (CofSD 1979). An effort to revitalize older, inner-city neighborhoods through another city program, City of Villages, was also in the development stages (GALLOWAY 2002).

The community's Vision Project garnered a better outcome than would have been achieved without the plan demonstrating that contrary to the media's representation, community members in what is continually described as an ethnically diverse, low-income community, were able to come together and develop a master plan, with an aim to foster economic growth, provide badly needed public services, and overcome a very real disamenity. Research pertaining to community participation suggests social and economic problems limit involvement from ethnic, racial and low-income groups, and that planners often perceive these groups as difficult to engage (BEEBEEJAUN 2006; FAGOTTO and FUNG 2006; LOWE 2008). A level of trust can be hard to achieve when speaking across cultures and languages and in a community like City Heights, resident's trust of the state could be limited by the lack of resource provision and history of poor planning practices. The Vision Project suggests an alternative reality; one of active participation. A network of activity focused on community improvement was able to enroll actors and stabilize around needs and projects.

The network was shaped by interactions with the state and economic decisions at various scales and although it stabilized around the freeway lid master plan, effectively enrolling the state, the state was more effective in shaping the outcome due to funding. None-the-less, tracing the actor-networks in City Heights

---

17 The redevelopment agency is a department of the city (CofSD 2003). It offers incentives to developers such as: assistance with site assembly, reduced development fees, permitting assistance, off-site improvements, housing programs, façade rebates, and low-cost financing incentives.

reveals an engaged community, active in their efforts to improve their neighbor-hoods and quality of life. The outcome reveals a power-differential, but not a community awaiting intervention. In fact, the Vision Project contains many of the design and material elements present in the master plan later developed by Price Charities, yet the role the Vision Project played in shaping later outcomes is not discussed in the post-Price era press.

City Heights Community Development Corporation

Network tracing can reveal numerous networks of activity with similar motiva-tions. Much of the pre-Price era improvement activity was facilitated through interaction with City Heights Community Development Corporation (CHCDC). Established in 1981, the CHCDC interacted with numerous community organiza-tions such as the community planning committee established in conjunction with the city's redevelopment area designation, the existing town council, two business improvement districts[18] and several active neighborhood associations[19] (POWELL, 2006). Community Development Corporations (CDCs) arose in the 1970s when federal funding for urban targeted renewal and social programs were experiencing cutbacks (FRISCH and SERVON 2006). Municipal revenues had long been directed towards the suburbs and CDCs took on the role of redirecting support back to the inner-city by forming relationships with government and non-profit organizations.

CDCs act as neighborhood advocates, creating and inserting "themselves into networks that are critical to their ability to build capacity" (90). For example, many CDCs engage in grassroots activism, organizing property clean-up and education and job training programs. They also enroll nonprofit housing agencies in an effort to increase affordable housing provision. In general, funding comes from a host of sources, including local governments, businesses, and non-profit organizations.

CDCs vary greatly in their activities from city to city (FRISCH and SERVON 2006). The CHCDC has sought a balance between community activism and affordable housing provision (POWELL 2006). They have been successful in reha-bilitating and managing a number of affordable complexes. Their efforts in activism have gained more attention, especially the annual International Village Celebration (IVC). Unlike other street festivals in the city focused food and craft booths, the IVC offered ethnic music, dance, and community services. During the 15-year tenure of the festival, the centerpiece of the IVC was a tent devoted to employment services (Figure 3). Surrounding 'Career Corner' were stages where members of the community performed music and dance from their home countries (Figure 4).

---

18   City Heights and El Cajon Boulevard Business Improvement Associations
19   Azalea Park, Cherokee Point, Fairmount Park, South Castle, Swan Canyon, Teralta Concerned Citizens, Teralta West (CD3 2010, 2003).

*Figure 3 (left): The employment services tent was the centerpiece of the
International Village Festival in City Heights. Photograph by author.*

*Figure 4 (right): A mariachi band performed at the International Village
Festival in City Heights in 2004. Photograph by author.*

Another activist effort organized through the CHCDC was a community garden. From 1990 to 1995, 1.5 acres, cleared in preparation for the I15 freeway extension, were converted to farming plots for 120 families for an annual lease of $10 per year (KUCHER 1995; POWELL 2006). The CHCDC subsidized the garden and organized monthly 'garden meetings' to talk about resources and area maintenance. In addition to providing community members with the opportunity to grow their own food, City Heights residents perceived the garden as a sanctuary; a safe place in the neighborhood.

Both the IVC and the community garden represented efforts by the CHCDC to provide access to supportive services for members of the community. The executive director of the CHCDC, along with a local neighborhood organizer cite another benefit to each of these activist efforts; a nurtured neighborhood spirit (KUCHER 1995; POWELL 2006). Community garden meetings facilitated communication across cultures because the CHCDC provided translators for numerous languages so that everyone would remain informed. In a similar vein, the IVC encouraged interaction through the experience of culturally diverse music, dance, and food. The executive director of the CHCDC likened the event to an "introduction" between neighbors and between neighborhoods in City Heights.

The CHCDC was also active in organizing and leading community clean-up efforts by enrolling funding from various sources. For example, the CHCDC helped community members procure $167,000 in 1995 as part of a Federal Enterprise Community grant for graffiti abatement and neighborhood clean-up (POWELL 1995). The presence of the CHCDC as an active member in City Heights' networks of improvement activity also shaped events such as the weekend retreat at a middle school where 400 attendees discussed the community's needs and issues[20]. City, state, and federal representatives, judges, police officers, teachers, ministers, and community members engaged in dialog, slept on cots in

---

20 Economics, community development, crime, land use, education, health and social services (YOUNG 1994).

the school's gym, and participated in day and nighttime field trips (YOUNG 1994a, 1994b, 1994c). Later the same year 200 people from the community attended a meeting in council chambers downtown to witness a city council vote on funding for the development of a police station in City Heights (YOUNG 1994b). They also presented council members with a list of recommendations from their brain-storming session held earlier in the year.

LAWLESSNESS, PASTICHE, AND PRICE: CONCLUSIONS ABOUT PHILANTHROPIC INTERVENTION

In the Price era press the standard article about City Heights begins by estab-lishing the need for intervention. A "reputation for high crime" (STETZ 2004) is suggested, and in both a celebratory and concerned tone, the community's diver-sity is illuminated. Residents are romantically fêted as "the threads of a multi-colored tapestry" (YOUNG 1994a), but media accounts also note the number of students in area schools that are eligible for "free or reduced-price meals" which is "a standard measure of low income" (RENO 2003). Price, is therefore lauded for comprehending the "potential amid the graffiti and rundown buildings" and fos-tering a revitalization plan that has "earned a national reputation as a model" for reversing blight in the inner city (PIERCE 2002).

> You have here a philanthropist who is beloved in this community (Maria Martinez-Cosio, Graduate Student, University of California, quoted in the film *Price of Renewal,* ESPINOSA 2006).

That the "major infusion of capital and caring from Price Charities" (RENO 2003) has had an impact on City Heights is not questioned. Whether the philanthropic organization's activities represent a positive model for replication in other com-munities, however, is a question that is unanswerable based on media represen-tations of processes of change in City Heights. Media standing which is vested in Sol Price and Price Charities obfuscates the roles of other actors. Some voices have simply been filtered out over time. For example, the voice of the CHCDC became quieter in the Price era press, though they remain active in the community fulfilling badly needed services.

The pre-Price activist community in City Heights is given limited credit for their role in identifying need and developing plans for neighborhood improve-ment. The dominant Price Charities discourse fails to examine interactions between this active community and the philanthropic organization. The role of important non-human actors such as historical planning documents and existing infrastructure remains uninterrogated as well. Table 1 lists numerous projects in which Price Charities was enrolled, either through planning or capital contribu-tions, or both. Many of the projects offer public services such as local police pro-tection, library and educational access, and recreational space. Within most U. S. communities these services are an expectation. Media accounts suggest through narrow depictions of a densely populated, diverse, and crime ridden community that City Heights was somehow not permitted the same expectations as most communities, overlooking the role the city plays in failing to equitably distribute municipal services. This limits the understanding of why philanthropic intervention is now necessary in inner-city communities.

The deeper analysis of the pre-Price era networks presented in this chapter demonstrates that residents and business owners in the community did have expectations preceding philanthropic intervention, and they actively sought to have these expectations fulfilled by the city. Community members and the CHCDC were actively seeking mobilization of public service goals through the Vision Plan. They had convinced the city of the need for a police station and the city had enrolled a site for development. Through continual community led efforts civic leaders were made aware of community issues, and they had been introduced to the community's active participation in neighborhood clean-up and self-policing.

Price Charities did not introduce the community to its needs. Nor did it introduce the city to the community's needs. Defining Price Charities as a catalyst is therefore, problematic. They neither initiated the vision for change, nor acted alone in development of that change. Price Charities was, in fact, one of many actors enrolled in networks focused on improving the quality of life in City Heights. Enrolled with the philanthropic organization were the city, county, numerous other non-profits, and most importantly, the community; supported through services, cultural awareness, and interaction with the city fostered through the City Heights Community Development Corporation.

Actor rather than Savior

Recent philanthropic enrollment in communities has been shaped by shifting planning paradigms such as the emergence of a humanistic and holistic approach to neighborhood planning (ABBOTT 1993), economic development strategies that recognize the retail potential in inner city neighborhoods (COYLE 2007), historical precedents that left inner-city communities lacking in standard services, and shifts in state support for municipalities. Price Charities' enrollment in City Heights' networks was set within the context of a state property tax proposition that greatly reduced municipal revenues and a national recession in the early 1990s. The city of San Diego, like the nation, was experiencing growth in its immigrant and low-income populations and exclusionary practices concentrated a large portion of this growing population in City Heights, placing further strain on what were already deficient services[21].

The context of community activism had been fostered by the CHCDC, which worked to mitigate constrains such as language and cultural difference; creating

---

21 Community members sought a moratorium on further higher density development in the community in the early 1980s, already recognizing that services were not meeting demand from the growing population (RISTINE 1984).

venues for interaction such as the festival and the community garden. Moreover, the CHCDC facilitated the needs of neighborhood organizations by garnering state and federal funds, and acted as a liaison between city departments and agencies, as well as other non-profit organizations. Although several media sources suggest Price Charities' enrollment was the catalyst of change[22], it was the community and the CHCDC at the forefront with established and active networks, which the philanthropic organization was able to enroll in rather than create.

Tracing the pre-Price networks of activity suggests that rather than leading the charge, Price Charities' capital acted to facilitate stability within already established networks that had preexisting goals for community improvement. This reinterpretation of processes of change in City Heights suggests a different role for Price Charities; rather than catalyst or savior, the philanthropic organization is an actor among many and through their interaction and the offer of capital reshaped differential power relations. Arguably, this is the philanthropic organizations' greatest impact, one that deserves further analysis.

> One of the most diverse communities in San Diego with more than 30 languages spoken, City Heights is a neighborhood on the rise. A business and residential area in the heart of Mid-City, City Heights is a true international village with amenities like the Mid-City Police Sub-station, a performance annex and several new public schools (Description of City Heights on Council District 3 Website 2010).

REFERENCES

ABBOTT, C. (1993): Five Downtown Strategies: Policy Discourse and Downtown Planning Since 1945. *Journal of Policy History* 5 (1): 5–27.

ALLEN, M. (2004): Mid-City Area Giving Firms a New Urban Option. *San Diego Business Journal*, 8 March, 6 & 31.

BEEBEEJAUN, Y. (2006): The Participation Trap: The limitations of Participation for Ethnic and Racial Groups. *International Planning Studies* 11 (1): 3–18.

BENNETT, K. (2008): In City Heights, a Housing Market of Juxtapositions. *Voice of San Diego*, 21 July, voiceofsandiego.org.

BIBERMAN, T. K. (2001): Dream of City Heights Improvement Becoming a Reality. *San Diego Daily Transcript*, April 20.

BICKERSTAFF, K., and J. AGYMAN (2009): Assembling Justice Spaces: The Scalar Politics of Environmental Justice. *Antipode* 41 (4): 781–806.

BUNNELL, G. (2002): San Diego, California: A Hundred-Year Planning Legacy (Periodically Interrupted). BUNNELL, G. (Ed.): *Making Places Special: Stories of Real Places Made Better by Planning*. Chicago, IL.

---

22 For example see ESPINOSA (2006); CLARK (2002); HERMAN (2002); RENO (2003); ALLEN (2004); BENNETT (2008); BIBERMAN (2001); ENSOR (2003); GEMBROWSKI (2001); MORGAN (2001); PIERCE (2002).

CAHILL, J. (2004): Conversational Interviews with Executive Vice President of Price Enterprises, January 2004 to November 2006 San Diego, CA.

CALLON, M. (1991): Techno-economic networks and irreversibility. LAW, J. (Ed.): *A Sociology of Monsters: Essays on Power, Technology, and Domination*. London , UK.

CANTLUPE, J. (1988): City officials, accused of running slum, to demolish 2 buildings. *San Diego Union Tribune*, 15 April, B3.

Council District 3 (2003): About City Heights. http://genesis.sannet.gov/infospc/templates/cd3/com_city_hts.jsp.

Council District 3 (2010): Community Organizations. http://www.sandiego.gov/citycouncil/cd3/communities/cityhts.shtml.

Census (1970): Decennial 1970 Census. Washington, D. C.: U. S. Census Bureau.

Census (1990): Decennial 1990 Census. Washington, D. C.: U. S. Census Bureau.

Census (2000): Decennial 2000 Census. Washington, D. C.: U. S. Census Bureau.

CLARK, CA. (2002): The Neighborhoods: City Heights. *San Diego Magazine*, May, www.sandiego magazine.com.

CLARK, CH. (2001): City Heights does its best to show its best. *San Diego Union Tribune*, 3 June, B2.

CLARK, CH., T. NICHOLS, R. LLOYD, K. K. WONG and P. JAIN (2002): Amenities Drive Urban Growth. *Journal of Urban Affairs* 24 (5):493–515.

CofSD. (1979): General Plan Update. San Diego, CA: City of San Diego.

CofSD. (2003): Redevelopment Agency Project Areas: City Heights. http://www.sannet.gov/redevelopment-agency/cityhts.shtml.

COLLUM, J. (2003): Conversational Interviews with CCDC Associate Project Manager. January 2003-December 2004, San Diego, CA.

COYLE, D. M. (2007): realizing the inner city. *Economic Development Journal* 6 (1):6–14.

DAVIS, M. (1992): Fortress Los Angeles: The Militarization of Urban Space. SORKIN, M. (Ed.): *Variations on a Theme Park: The New American City and the End of Public Space*. New York, NY.

DUNCAN, J. and D. LEY (1993): Introduction: Representing the place of culture. DUNCAN, J. and D. LEY (Eds.): *place/culture/representation*. New York, NY.

ENSOR, D. (2003): Village hailed as 'smart growth'. *San Diego Union Tribune*, 22 February, B2.

ESPINOSA, P. (2006): The Price of Renewal. ESPINOSA, P. (Ed.): *California and the American Dream*. United States: Beyond the Dream LLC (Documentray).

FAGOTTO, E. and A. FUNG (2006): Empowered Particiaption in Urban Governance: The Minneapolis Neighborhood Revitalization Program. *International Journal of Urban and Regional Research* 30 (3): 638–655.

FORD, L. and E. GRIFFIN (1979): The Ghettoization of Paradise. *The Geographical Review* 69 (3): 140–158.

FORD, L. R. (2005): *Metropolitan San Diego*. Philadelphia, PA.

FREEMAN, L. (2005): Displacement or Succession? Residential mobility in Gentrifying Neighborhoods. *Urban Affairs Review* 40 (4): 463–491.

FRISCH, M. and L. J. SERVON (2006): CDCs and the Changing Context for Urban Community Development: A Review of the Field and the Environment. *Community Development* 34 (7): 88–108.

GALLOWAY, T. (2002): Interview with City of San Diego Associate Planner-City of Villages Pilot Program. 15 November, San Diego, CA.

GEMBROWSKI, S. (2001): City Heights retail site opens. *San Diego Union Tribune*, 10 November, http://www.signonsandiego.com/.

GRAHAM, J. (2005): Conversational Interviews with San Diego Police Department Community Outreach Officer. July 2002 to July 2005, San Diego, CA.

GREEN, K. (2005): World's on view on this avenue. *San Diego Union Tribune*, 6 September, B1.

GROGAN, P. S. and T. PROSCIO (2000): *Comeback Cities: a Blueprint for Urban Neighborhood Revival.* Boulder, CO, USA.

HEINE, G. (1973): University Avenue. In *Images of San Diego Neighborhoods*, edited by L. Ford and B. R. OBrien. San Diego.

HENSLEY, H. C. (1956): Steam Trains to East San Diego. *Journal of San Diego History* 2 (1).

HERMAN, K. (2002): A Slow and Steady Revival in San Diego. *Blueprint Magazine*, 22 January, 17.

HILLIER, J. (1997): Values, Images, Identities: Cultural Influences in Public Participation. *Geography Research Forum* 17: 18–36.

JAHN, E. (1992): Council buries ambitious plan for I-15 cover. *San Diego Union Tribune*, 28 April, B 3 & B5.

JOHNSON, S. (2006): Conversational Interviews with City of San Diego's Community Services Division Homeless Services Administrator. March 2005-July 2006, San Diego, CA.

KILLROY, C. (1993): Temporary Suburbs: The Lost Opportunity of San Diego's National Defense Housing Projects. *The Journal of San Diego History* 39 Winter-Spring (1–2).

KUCHER, K. (1991): Congressmen seek funds to develop park on covered freeway at 40th St. *San Diego Union Tribune*, 31 May, B7.

KUCHER, K. (1995): Green thumbs turn blue. *San Diego Union Tribune*, 15 December, B1.

LATOUR, B. (2005): *Reassembling the Social: An Introduction to Actor-Network Theory.* New York, New York.

LAW, J. (1992): Notes on the Theory of the Actor-Network: Ordering, Strategy, and Heterogeneity. *Systems Practice* 5 (4): 379–393.

LEES, L., T. SLATER and E. WYLY (2007): *Gentrification.* New York, NY.

LOWE, J. (2008): Limitations of commuity development partnerships: Cleveland Ohio and Neighborhood Progress Inc. *Cities* 25: 37–44.

MATTINGLY, D. (2001): Place, teenagers and representations: lessons from a community theatre project. *Social and Cultural Geography* 2 (4):445–459.

MORGAN, N. (2001): Sol Price, at 86, helps make City Heights a pioneer. *San Diego Union Tribune*, 7 December, http://www.signonsandiego.com/.

MORRIS, E. (2001): William D. Jones (award recipient). *San Diego Dialogue*, http://www.sandiego dialogue.org/events/livablebios.html.

MURDOCH, J. (1997): Towards a geography of heterogeneous associations. *Progress in Human Geography* 21 (3): 321–337.

PIERCE, E. (2002): On to new Heights. *San Diego Union Tribune*, 21 April, I–1 and I–4.

POWELL, J. (2006): Conversational Interviews with Executive Director, City Heights Community Development Corporation, June 2004-November 2006. San Diego, CA.

POWELL, R. W. (1995): U. S. gives city neighborhoods $3 million for revitalization. *San Diego Union Tribune*, 29 June, B1.

PRED, A. R. (1984): Place as Historically Contingent Process: Structuration and the Time-Geography of Becoming Places. *Annals of the Association of American Geographers* 72 (2): 279–297.

Price (2002): Crime in City Heights. San Diego.

Price (2009): *Price Charities* [Web Page]. Price Charities 2009. Available from http://www.price charities.com/.

RENO, J. (2003): The Sucess of Price. *San Diego Magazine*, April, 29–32.

RICHMOND, M. (1991): City Heights revitalization backed by council, residents. *San Diego Union Tribune*, 31 May, B7.

RISTINE, J. (1984): Building moratorium studies for Mid-City. *San Diego Union Tribune*, 6 March, B3.

ROTH, A., L. VANDER HAAR and E. L. VANDER HAAR (2006): Media Standing of Urban Parkland Movements: The Case of Los Angeles' Taylor Yard, 1985–2001. *City & Community* 5 (2): 129–151.

228 Brenda Kayzar

ROYAL, D. (2005): Conversational Interviews with CCDC Senior Projects Manager, Affordable Housing. November 2001-May 2005, San Diego, CA.

RYPKEMA, D. D. (2003): The Importance of Downtown in the 21st Century. *American Planning Association Journal* 69 (1): 9–15.

Sandag (2003): Census 2000 Profile: City Heights Community Planning Area. San Diego, CA: San Diego Associaiton of Governments.

Sdhc (2009): Affordable Housing Resources. San Diego, CA: San Diego Housing Commission.

Sdusd (2009): School Profiles: 2007: Issued 2007 for schools open in academic year 2005–06 San Diego, CA: San Diego Unified School District.

SHOWLEY, R. M. (1999): *Perfecting Paradise*. Carlsbad, CA: Heritage Media Corp.

SMITH, N. and P. WILLIAMS (Eds.) (1986): *Gentrification of the City*. Boston.

SMITH, R. G. (2003): World city actor-networks. *Progress in Human Geography* 27 (1): 25–44.

STEPNER, M. (2007): Conversational Interviews with Former City of San Diego Planning Department Director and City Architect, Professor and Board Member, New School of Architecture. August 2003-July 2007, San Diego, CA.

STETZ, M. (2004): Festival celebrates residents' diversity. *San Diego Union Tribune*, 6 June, B1–2.

TOSCANO, R. (2009): Cheap Homes Still Flying Off the Shelves. *Voice of San Diego*, 10 June, voiceofsandiego.org.

TUREGANO, P. (1985): Lenders' reluctance may speed blight, study says. *San Diego Union Tribune*, 7 March, B3.

WEISBERG, L. (1991): Plans for town center atop freeway advance. *San Diego Union Tribune*, 22 May, B1.

YOUNG, E. (1994a): City Heights stives to cut crime, boost business. *San Diego Union Tribune*, 26 February, B1 & B7.

YOUNG, E. (1994b): Forum seeks a vision for City Heights. *San Diego Union Tribune*, 23 April, B1.

YOUNG, E. (1994c): Police-station plans gain in City Heights. *San Diego Union Tribune*, 12 November, B1.

ZUKIN, S. (1995): *The Culture of Cities*. Cambridge, MA.

Michaela Benson / Ed Jackiewicz

QUESTIONING 'THE RIGHT TO STAY PUT' AS AN EXPLANATION OF THE IMPACT OF LIFESTYLE MIGRATION AND RESIDENTIAL TOURISM DEVELOPMENT

INTRODUCTION

This chapter explores the prospects of 'the right to stay put' as a concept for explaining the impact(s) of lifestyle migration and residential tourism development. With an empirical focus on such migrations and developments in Panama, it questions whether 'the right to stay put' can adequately account for the various impacts felt on a local level. Although there is some evidence of displacement coinciding with the recent in-migration of North American and European populations, understanding this phenomenon within the wider context of social and economic change in Panama reveals that the reality is far more complex than 'the right to stay put' paradigm allows.

The right to stay put has become increasingly elusive to many around the globe. The fact remains that most people do not *want* to move and as HARTMAN (1984) states this is often a "severely damaging experience" (302). Most would prefer to stay close to family, friends, and familiar surroundings if given the choice. Of course, mobility has significantly different meanings when one compares someone who voluntarily moves to a nicer location, for a better job, etc. with one who moves because they have no reasonable alternative when the place they know as home has been altered by economic, environmental or other circumstances. In this chapter, we examine the issue of the right to stay put in the country of Panama. More precisely, we look at how the increasing presence of lifestyle migrants, i.e. voluntary migrants, in Panama has impacted local individuals and their ability to stay put, i.e. involuntary migrants.

Lifestyle migration is defined as the "spatial mobility of relatively affluent individuals of all ages moving either part-time or full-time to places that are meaningful because, for various reasons, they offer the potential for a better quality of life" (O'REILLY and BENSON 2009, 2). During the last decade, Panama has emerged as one of the world's leading destinations for lifestyle migrants due to several factors discussed below (see also JACKIEWICZ and CRAINE 2010). Another important contributing factor to the popularity of Panama is how it is presented (i.e. sold) in the media to potential lifestyle migrants. Of most importance to this chapter however, is how the increasing popularity of Panama is altering the reality for Panamanians. Before moving on to that issue, it is first essential to provide an overview of why Panama has become such a popular destination, followed by a discussion of lifestyle migration/migrants. We then take a closer look at specific locations in Panama that are popular for lifestyle migrants and how the locals are responding to these spatial transformations.

THE SELLING OF PANAMA

Panama has emerged as a leading destination among lifestyle migrants for several important reasons, which include, among others, a favorable tax environment, dollar economy, relatively low cost of living, proximity to US (the country of origin of many of the migrants), quality healthcare, large percentage of English speakers (especially when compared to most of its Central/South American neighbors), tropical climate. However, what has been arguably the biggest force driving this movement is how Panama has been marketed through various media outlets.

Upon arrival in Panama, one is immediately bombarded with billboards and magazines promoting the local real estate (see figure 1). However, the most prolific media in bringing buyer and seller together is, not surprisingly, the Internet. It is hard to imagine that this industry would be nearly as successful without online promotions and searches. If one has an interest in living or investing in real estate in a foreign country, detailed information is immediately attainable and Panama often emerges as a desirable locale.

If we look more closely at commentary taken from the Internet, we can better understand how Panama is being portrayed and sold to a global audience. This first excerpt is from the *Panama Offshore* website:

Many people are coming to Panama in search of a cheap piece of land so they can build their dream home in paradise.

All of these projects are unique in their own way, but what they all have in common, is that they are all first-world quality developments with first-class facilities and first-class security systems, working 24 hours a day to protect the resident.

And this one is from International Living, a leading promoter of foreign real estate, a publication that many North Americans living in Panama stress that they had turned to when they had first considered moving abroad (McWatters 2008):

Panama is also far more developed and modern than most people anticipate. The idea that moving to Panama means a move away from the First World could not be farther from the truth. People who move to Panama can typically find the same luxuries and amenities that they would hope for anywhere else, and can even afford more of them due to Panama's low cost of living.

There are several noteworthy points made evident in these brief excerpts. The first is that you are being offered an opportunity to move to 'paradise' that sends a powerful signal to the would-be buyer (see Osbaldiston 2012 for a more detailed discussion). Individuals can then begin to imagine themselves lying in a hammock, surrounded by turquoise waters and white sand and sipping on a cool beverage; reminiscent to how Florida was sold to northern Americans forty or fifty years ago.

It is also important for the marketers to erase any potential apprehensions that potential buyers might feel. Safety is of course near the top of that list. Latin America in general is still perceived by many to be dangerous and crime-ridden. In the first excerpt, "security" and "24 hour protection" are highlighted to assuage those concerns. Others still perceive Panama (and all of Latin America) as the "Third World" that conjures up many not so pleasant images that would discourage potential investors. Both excerpts address this issue "head on" by emphasizing luxuries, amenities, technology, etc. and even evoking a First World lifestyle.

This marketing campaign has obviously worked as the numbers of people moving to Panama continues to increase creating many opportunities and challenges for the local population. However, more relevant to this paper is the subsequent movement instigated by the arrival of foreigners. First, while precise numbers of new arrivals is difficult to ascertain, there are many originating from the US, Canada, Europe and other Latin American countries.[1] Many of these are considered wealthy or at least middle-class by industrialized country standards. They have the financial resources to relocate, purchase or lease a home, and live quite comfortably by local standards. This movement has precipitated the construction of homes, many in brand new developments around the country. The need for labor has led to the migration of many workers from within and outside of Panama into the capital city or wherever the construction is occurring. These jobs typically pay a decent wage, but are often not long term. Once the construction is complete, there is the demand for laborers to maintain homes and yards and provide other services for the expatriates. Thus, this has set off a second wave of migration to destinations with relatively large numbers of foreigners.

The government in promoting this type of growth has designated certain areas around the country as tax havens for these investors/migrants. This includes the capital city but also other popular destinations such as the mountain community of Boquete and the Caribbean archipelago of Bocas del Toros (the latter two are used as examples later on). These developments have stirred up migration from within Panama as well as from neighboring countries. One insidious aspect of this worth mentioning is the movement of prostitutes from Colombia and other nearby countries, largely encouraged by the growing presence of these new relatively well-to-do residents.

In turn, what we attempt to illustrate here is that by encouraging lifestyle migrants to choose Panama as their place of residence, other movements are set in motion that challenge locals' right to stay put. Understandably, in the absence of other economic opportunities, many will choose to migrate to where the new

---

1   This effort is not aided by the fact that the US government do not have accurate estimates of the numbers of their citizens living abroad; they are not counted on the census due to the cost of conducting such an exercise abroad.

development is occurring with the hope of finding gainful employment and new opportunities. Whether or not this is the right decision, especially in the long run, is yet to be determined. This chapter proceeds by taking a more in-depth look at lifestyle migration followed by a closer examination of how Panamanians, both in Boquete and Bocas del Toro, variously experience living in these areas of notable foreign populations.

LIFESTYLE MIGRATION: AN OVERVIEW AND CONSIDERATION OF IMPACTS

The movement of such affluent individuals into certain areas of Panama can broadly be defined as lifestyle migration, a phrase intended to draw attention to the relative affluence of these migrants (in comparison to other migrants) and to highlight the desire for a better way of life as motivating migration in the first place (BENSON and O'REILLY 2009; O'REILLY and BENSON 2009). This is a social phenomenon that has been taking place around the globe, as people move to and from an increasing number of destinations. In many respects, this is a form of migration that has often been overlooked in migration research.

The literature on lifestyle migration has focused on four themes: the decision to migrate, everyday lives within the destination, relationships with the local population, invariably framed in terms of integration, and, to a lesser extent, the impact of these migrations and the social change that they incur. It is the focus on the decision to migrate that has produced many of the previous conceptualizations of this social phenomenon, focusing the driving forces behind migration with little reflection on how these then translate into the everyday lives led within the destination. Such migration has therefore been described as international retirement migration (IRM), reflecting the coincidence of migration with this particular stage in the life course (see for example RODRÍGUEZ, FERNÁNDEZ-MAYORALAS and ROJO 1998; KING et al. 2000; OLIVER 2008); international counterurbanization, emphasizing the antimodern and antiurban longings that lie at the root of this form of migration (BULLER and HOGGART 1994); and residential tourism which draws attention to the roots of this migration trend and its similarities to tourism (see for example RODRÍGUEZ 2001; MCWATTERS 2008).

As BENSON and O'REILLY (2009) argue, none of these prior conceptualizations adequately capture the full nature of this migration trend. IRM neglects the other, younger migrants who are also part of this trend, while international counterurbanization overlooks the relationship between antiurban/antimodern desires and everyday life within the destination, while also excluding urban longings from the equation. Similarly, the reduction of migration to tourism, as residential tourism promotes, undermines the diverse motivations and experiences of these migrants and has yet to provide an explanation of why people would choose to turn their tourism experiences into a way of life. In all cases, migration represents a turning point in the migrants' lives, the point at which they can realize their longings for a better way of life. In contrast, lifestyle migration is intended to capture the dynamics not only of the decision to migrate but also the location of this within the individuals' wider biography and their lifestyle trajectory, incorporating their lifestyle choices before and after migration (BENSON and O'REILLY 2009; O'REILLY and BENSON 2009). Migration thus is not such a severe rupture from life before migration, just one step en route to a better way of life (BENSON 2011).

While there is a increasing body of researchers exploring the various articulations of this phenomenon as evidenced by the lifestyle migration hub (www.uta.fi/yky/lifestylemigration/index.html), very little attention has been given to how this is experienced on a local level (for a notable exception see MANTECÓN 2008), by the people in the communities hosting these migrants, and wider changes it spurs. However, as we argue here, lifestyle migration can have very real consequences for the local population, bringing about infrastructural problems as we have seen in the case of Spain and significant social and environmental change (HUETE, MANTECÓN and MAZON 2008; MANTECÓN 2008). In part, this research in Spain reflects its longstanding status as perhaps the prototypical site as a lifestyle migration destination, with an established British population, documented by O'REILLY (2000) and OLIVER (2008), but also a reputation as a retirement destination (for a review of this literature see RODRÍGUEZ et al. 2005; CASADO-DIAZ 2006). Although the impacts of such extensive in-migration and unprecedented development emerge from particular local and national contexts, there are important lessons to be learned from this example which can perhaps help to shape discussion about the possible outcomes of this form of migration.

As HUETE, MANTECÓN and MAZON (2005) argue, residential tourism development in Spain has had a variety of outcomes. In particular, the economic growth spurred by this phenomenon has been largely positively received, helping with the overall development of areas concerned. Indeed, it is this aspect of residential tourism development that many governments, including Panama, have embraced and used as justification for implementing large-scale development plans. In Spain, a perhaps unanticipated side effect of this has been an influx of younger people, attracted by the job opportunities created by tourism and development. In this respect, lifestyle migration has contributed to a wider trend of demographic growth in certain areas. Nevertheless, there have also been negative consequences, attributed largely to the lack of sustainability of development plans, including a lack of public infrastructure, as well as question marks over whether the economy in the area will continue to flourish once the incoming population slows, and the possibility of longer term damage to the image of Spain as a tourism destination. These consequences have taken a particular shape because of the trajectory spanning from tourism to residential tourism development and lifestyle migration. Indeed as RODRÍGUEZ (2001) has argued, tourism is a recruiting post for migration of this type; and this is twofold, with places marketed on the basis of their existing status as tourism destinations, and with people choosing destinations on the basis of their holiday experiences.

Turning towards Central America, the main focus of researchers has been on Costa Rica – perhaps the quintessential site of lifestyle migration in this part of the world. Recent research by authors BARRANTES-REYNOLDS (2011) and VAN NOORLOOS (2011a and 2011b), examine the social impacts of real estate development for local actors. As VAN NOORLOOS (2011b) argues, there is a very real sense in which such local actors, through the in-migration of foreigners and related developments, experience alienation as land is increasingly privatized (see also McWATTERS 2008 in the case of Panama). At the same time, this in-migration spearheads the circulation of money, goods and people between the centre and peripheral areas, with people moving to the area to supply labor to these incoming populations. While not yet as developed as in the case of the British in Spain, it is understandable that concerns over the economic, social and environmental impacts of lifestyle migration in Central America need to be taken seriously.

LIFESTYLE MIGRATION IN PANAMA

In Boquete, a town struggling to hold its position in an unstable global coffee market, residential tourism is one option that has been explored to maintain a position on the world stage. The coffee crisis of the late 1990s and early part of this century affected Boquete, which is a small producer, particularly badly. Alongside other destinations, Boquete was chosen by the government as a zone where any new properties would be exempt from property tax for twenty years. Combined with special visas aimed at bringing *jubilados* (pensioners) into the country, particularly North Americans, this signaled the efforts of the Panamanian government to increase their Foreign Direct Income through property development.

The rise of Boquete as a retirement destination occurred just after the collapse of the coffee market, and arguably, was a product of this. The cost of land was very low because it was no longer valuable agricultural production land. At the same time, there were people who, offered money for their land and homes, chose to sell up and move out of Boquete. However, much of the land which has now been converted into residential plots was once coffee plantations. For example, Valle Escondido, one of the most well-known residential tourism developments in Boquete, is built on the site of a former coffee plantation. Other coffee producers have sold off small parcels of their land to private buyers (rather than developers), with the result that many incomers actually lay claim to a small coffee finca.

Undeniably, lifestyle migration into Boquete has brought economic advantages. Very simply, planning applications have significantly boosted the municipal income and thus there has been more money available to fix roads and renovate the local park for example. But there is a downside to this, as many younger members of Panamanian population find that they can no longer afford to buy houses, with land prices inflated as a result of residential tourism development (in the early days of such development land cost just under $3/m^2$, but now, can rise up to $80/m^2$), and the cost of groceries in the supermarkets rising. These aspects of economic change have had a negative impact on the quality of life available to the locally-resident Panamanian and indigenous population. But such impacts need to be considered within the context of broader changes taking place in the area.

For example, another result of the coffee crisis is that there are fewer jobs for the indigenous (Ngäbe Bugle) labour force.[2] However, some of these have found jobs working as gardeners or maintaining the small coffee plantations owned by the incoming population. In addition, many of the homes hire housekeepers. In this respect, although it by no means compensates for the number of jobs lost in the area, it appears that there are new opportunities for local people that are particular to the needs of the incoming population. In addition, it is apparent that some younger Panamanians have moved from Panama City to Boquete because of the opportunities that it offers. For example, Habla Ya, a successful Spanish language school was set up in Boquete by Sergio Díaz and Julio Santamaría from Panama City, in 2005. It has since expanded to include the adventure tour company Explore Ya, and a second school in Bocas del Toro. When they first sent up the school, Diaz and Santamaria had hoped to capitalize on the needs of the incoming foreign population, many of whom did not speak Spanish, as well as tourists. The successful marketing of Habla Ya and the experiences that it offers – including adventure tourism – has no doubt played a role in the further development of tourism in Boquete. Finally, it is certainly the case that there are more jobs in construction than ever before, which have perhaps attracted workers from David, as well as attracting local people who have tried to re-train. Visibly, in the last few years, not only has there been the development of property for the incoming lifestyle migrant population, there are larger numbers of tourists in the streets and a larger number of businesses aimed at catering to their needs.

Although it is difficult to trace exactly the changes in Boquete, it is clear that in the last few years, possibly in part influenced by the successful marketing of Boquete as a retirement destination, tourism has flourished. Since 2007, when one of us (Benson) first visited Boquete, the presence of young travelers from around the world has become increasingly noticeable as a result of the type of tourism the area is promoted as offering. Habla Ya recently prepared a press release, stressing that the retirement haven was increasingly becoming an 'Eco-adventure Capital'. While real estate development has noticeably slowed as a result of the financial crisis, backpacker tourism and eco-tourism have become a visible presence in the town. In 2010, there were 7 hostels and a number of tourism companies – offering tours of the coffee plantations, ziplining, hiking, rafting, trips out to the islands and the national marine park surrounding Isla Coiba – in the town of Alto Boquete. Such enterprises are run both by incomers (lifestyle migrants and internal migrants) and local entrepreneurs.

---

2   Panama has a significant indigenous population made up of seven tribes. The Ngäbe Bugle are by far the largest of these and have traditionally provided a 'migrant' labour force in Boquete and more widely in the region of Chiriqui (where Boquete is located), an area noted for its agricultural production.

We can only provide a brief sketch here of the different ways in which life-style migration into the area is responded to by the local Panamanian population. On the one hand, the outline presented demponstrates the need to consider wider structural changes and context in understandings of residential tourism development. On the other hand, it reveals that the incoming population does not always result in displacement of the existing residents as new job opportunities may become available. Furthermore, it demonstrates that the local is dynamic and constantly undergoes (often unpredictable) change.

Boquete is a unique case, perhaps unparalleled in the rest of Panama. As a brief point of comparison, Bocas del Toro has also witnessed some changes in recent years, which perhaps provide further insights into the use of 'right to stay put' within the broader Panamanian context. STEPHENS (2008) recounts this island's long relationship with global trade as a result of its historic position in the banana industry – the Hotel Grande Bahia in Bocas del Toro town is housed in the building that used to be the headquarters of the United Fruit Company – with banana trade starting in the Chiriqui lagoon in the late 1880s. The prospect of banana trade had brought many Europeans and North Americans to the area at the end of the 19th century, which is attested to today by the number of foreign graves in the cemetery. Attracted by the prospect of work at this time, many workers from other Caribbean locations moved to Bocas del Toro at the beginning of the twentieth century. The turbulent history of the banana trade, its response to disease and labor unions, was played out in Bocas del Toro throughout the twentieth century, and the supply of bananas from Central America started to dwindle.

Today its role in the economic market has shifted, as it plays host to many independent travelers and foreign residents attracted by the relaxed way of life that it offers. Once the headquarters of the United Fruit Company, it is now a popular spot with backpackers from around the world, and has attracted a number of lifestyle migrants, some of whom run businesses. Similarly, the change in population has also brought other migrants into the area, most noticeably, Hispanic taxi drivers migrating from other parts of Panama to take up opportunities in Bocas del Toro.

What is significant in the case of Bocas del Toro, and is presented here as only a brief example of the consequences of land-value increase, is the conflict over land that has been going on in recent years. This was recently evocatively portrayed in Anayansi PRADO's (2011) film *Paraiso en Venta* (Paradise for Sale). The heated debate around land rights in this part of Panama demonstrates, we argue, another side to 'the right to stay put'. Precipitated by the sudden increase in land value and the laws distinguishing right of possession from titled land – which is extremely common on coastal land in Panama, and where the government owns the land – public debate and advocacy most often focus on local people fighting to stay on land that they claim their families have occupied for generations. Indeed, STEPHENS summarizes this conflict over land, stressing that '[q]uestionable property ownership and doubtful legal titles have caused numerous boundary and property disputes throughout the islands' (2008, 79).

Such conflicts are regularly reported in the expatriate newspapers, the local *Bocas Breeze*, and the Panama City based, *Panama News*. The coverage demonstrates that these conflicts extend also to members of the expatriate population. For example, the owners of the botanical gardens, *Finca Los Monos*, recently found themselves caught up in a dispute over the ownership of their land, which has turned into a long and difficult legal battle. Members of the local community, whether longstanding residents or newly-arrived lifestyle migrants, argue that the government, or other elite Panamanians and occasionally international property developers, are trying to steal their land from them to make more money (of which they will not see a dime), the result perhaps, of the massive inflation of property prices on the archipelago over the last few years.

The fact that these disputes appear to have affected people in a range of positions within the community demonstrates the scale of the problem. And while it is evident that the recognition of Bocas del Toro as a tourism destination and as a lifestyle migration destination has had a causal effect on the price of land, the anger does not appear to be directed towards recent incomers. It is rather the case, as one respondent claimed, '[w]e only don't like people who cheat and steal'.

While we have not had the opportunity to delve any further into this case here, it is obvious that it needs further consideration, particularly in light of the complexities of the legal system in Panama, and the changing legislation regarding property ownership brought in by Martinelli, the current president. Nevertheless, it is clear that there is an overwhelming sentiment at play which is that people have the 'right to stay put' on the land which their families have occupied for generations, just as those who have legitimately purchased their properties should also resist conflicting claims. The fact that the claimed villains of these stories, those trying to take people's land, are often large conglomerates or the elite is perhaps revealing of local resistance to globalization and the commoditization of the locality.

CONCLUSION

As we have argued in this chapter, lifestyle migration and related developments may have diverse impacts, and are responded to in a variety of ways by local actors; while it is undeniable that the in-migration of North Americans and Europeans has coincided with a certain degree of displacement of Panamanian and indigenous populations, it is important to be wary of the extent to which demand for property and land alone drives this displacement. The empirical evidence that we have presented in this chapter demonstrates that different local contexts, in particular the relationship of places to the global, produce a complex dynamic, in which recent incomers and property development are one factor. In this respect, presenting a direct correlation between in-migration into these communities and displacement can mask wider structural changes that may also contribute to the displacement of resident populations.

In our increasingly hypermobile society, staying put is increasingly less desirable and movement is, at times, the best option for many seeking to improve their lives. Panama is literally a "crossroads" destination with many people moving in to, out of and around the country. Movement in to the country is being encouraged from the government level; other structural forces, largely economic, are encouraging other migrations that are fundamentally altering the landscape.

In this paper, we give brief descriptions of change occurring in two destinations: Boquete and Bocas del Toro simply to illustrate the great diversity of experience. The diversity ranges from why these destinations have become popular, to who chooses to move there, and who among the locals decides to stay (or leave). Whether we view those who leave as losing their right to stay put, we argue, is not so easily answered.

ACKNOWLEDGEMENTS

The empirical data reported here in respect to Boquete and Bocas del Toro was collected by Michaela Benson as part of a project 'Contemporary North American Migration to Panama: Expectations, Outcomes and Dynamics' funded by the British Academy (Grant Number SG-53957).

REFERENCES

BARRANTES-REYNOLDS, M-P. (2011): The Expansion of 'Real Estate Tourism' in Costal Areas: Its Behaviour and Implications. *Recreation and Society in Africa, Asia and Latin America* 2(1): 50–67.

BENSON, M. (2011): *The British in Rural France: Lifestyle Migration and the Ongoing Quest for a Better Way of Life.* Manchester.

BENSON, M. and K. O'REILLY (2009): Migration and the Search for a Better Way of Life: a Critical Exploration of Lifestyle Migration. *The Sociological Review* 57 (4), 608–625.

BULLER, H. and K. HOGGART (1994): *International Counterurbanization.* Ashgate: Brookfield, VT.

CASADO-DÍAZ, M. Á (2006): Retiring to Spain: An Analysis of Difference among North European Nationals. *Journal of Ethnic and Migration Studies*, 32(8), 1321–1339.

HARTMAN, C. (1984): The Right to Stay Put. GEISLER, C. C. and E. J. POPPER (Eds.): *Land Reform, American Style.* Rowman and Allanheld: Totowa, N.J.

HUETE, R., A. MANTECÓN and T. MAZON (2008): Analysing the Social Perception of Residential Tourism Development. COSTA, C. and P. CRAVO (Eds.): *Advances in Tourism Research.* Aveiro: IASK, 153–161.

JACKIEWICZ, E. and J. CRAINE. (2010): Destination Panama: An Examination of the Migration-Tourism-Foreign Investment Nexus. *Recreation and Society in Africa, Asia and Latin America.* 1(1), 5–29.

KING, R., A. M. WARNES and A. M. WILLIAMS (2000): *Sunset Lives: British Retirement to Southern Europe.* Oxford.

MANTECÓN, A. (2008): a experiencia del turismo. Un estudio sociológico sobre el proceso turístico-residencial (Barcelona: Icaria).

McWATTERS, M. R. (2008): *Residential Tourism: (De)Constructing Paradise*. Bristol.

OSBALDISTON, N. (2012): *The Great Urban Escape: Seeking Authenticity in Place, Culture and Self*. London.

O'REILLY, K. (2000): *The British on the Costa del Sol*. London.

O'REILLY, K. and M. BENSON (2009): Lifestyle Migration: Escaping to the Good Life. BENSON, M. C. and K. O'REILLY (Eds.): *Lifestyle Migration: Expectations, Aspirations and Experiences*. Farnham: Ashgate, 1–14.

OLIVER, C. (2008): *Retirement Migration: Paradoxes of Ageing*. London.

PRADO, A. (2011): *Paraiso en Venta* (film).

RODRÍGUEZ, V. (2001): Tourism as a recruiting post for retirement migration. *Tourism Geographies* 3 (1), 52–63.

RODRIGUEZ, V., G. FERNANDEZ-MAYORALAS, M. A. CASADO and A. HUBER (2005): Una perspectiva actual de la migración internacional de jubilados a España. Rodriguez, V., M.A. Casado and A. Huber. *La migración de jubilados europeos en España*. Servicio de Publicaciones, CSIC, 15–45

RODRÍGUEZ, V.; G. FERNÁNDEZ-MAYORALAS and F. ROJO (1998): European retired in the Costa del Sol: a cross-national comparison. *International Journal of Population Geography*, 4 (2), 183–200.

STEPHENS, C. (2008): *Outline of history of the province of Bocas del Toro, Panama*, Florida.

VAN NOORLOOS, F. (2011a): A Transnational Networked Space: Tracing Residential Tourism and its Multi-local Implications in Costa Rica. *International Development Planning Review* 33 (4), 430–444.

VAN NOORLOOS, F. (2011b): Residential Tourism Causing Land Privatization and Alienation: New Pressures on Costa Rica's Coasts. *Development* 54 (1), 85–90.

Jason Dittmer

RACE, DISPLACEMENT, AND THE AMERICAN COMICS INDUSTRY'S 'RELEVANCE' MOVEMENT

INTRODUCTION

This brief chapter addresses the ways in which race inflects both representations of displacement and gentrification in American popular culture and also the ways in which a 'right to stay put' might be implemented in the United States. In doing so I hope to highlight the potential for inequality should such a right be theorized and implemented strictly on the basis of a neo-Marxist understanding of class. This chapter begins with a review of the recent 'right to stay put' literature with a focus on the way in which race has been considered. The chapter then shifts gears to a specific representation of African-American displacement and gentrification from the pages of *Captain America* in the early 1970s. This narrative, a reader's response to it, and the editor-writer's reply are then used to inform a conclusion which calls for openness to new thinking on residence and 'staying put' that remains attentive to the significance of race in the historical evolution of the U.S. housing system.

RACE AND THE RIGHT TO STAY PUT

Chester HARTMAN's (1984) 'right to stay put' has experienced a resurgence of interest in recent years, owing partly to the twenty-fifth anniversary of his seminal publication, and also owing to its relevance as a tool in combating the predations of the current neo-liberal moment in urban governance, which has valorized gentrification as a model of social integration and improvement. It is perhaps indicative that this right is best developed in the social democracies of Scandinavia (IMBROSCIO 2004) and has become something of a *cause celebre* among progressive scholars and activists in the United States, the United Kingdom, and elsewhere. The right to stay put does however have its critics, who question the concept's grounding in a liberal notion of rights (e.g. GRAY 2009).

Indeed, the flattening and universalizing qualities of liberalism have an impact on any proposed right to stay put. Such a right would undoubtedly be uneven in its application:

> A citizen's right to fair treatment in the judicial system, to participate in the democratic process, and to educational opportunities or social services – while ostensibly absolute and universal – all remain highly imperfect and impure in the real world, especially because of defects and dilutions rooted in class, race, and gender dynamics (IMBROSCIO 2004, 581).

Given the traditional focus of Marxist analysis on class, questions of displacement and gentrification have usually focused on the impact of gentrification on the poor, despite its clear connections to race, especially in the context of the United States:

> The housing question touches deeply on issues of race and racism. A right to decent, affordable housing inevitably will involve a far greater degree of residential integration than now is the case. Major resistance to dealing with the fundamental flaws in the nation's housing system may stem from society's resistance to dealing with race issues (HARTMAN 2003, 156).

Thus, a focus on liberal rights requires rigorous attention to the way in which differential outcomes occur under universalizing pressures, especially when it comes to race in the United States.

Race has figured significantly in the new 'right to stay put' literature, with MANZO et al. (2008) arguing that displacement from public housing threatened the significant gains that have been made at racial integration in such a setting. SLATER (2008, 218) has argued that "[i]n studies of gentrification and the intersections of race and class, the racial identities of those involved must not distract attention from structural inequalities in society, of which gentrification is the neighbourhood expression." Still, despite this prioritization of class over race, he opens a space for interrogating the role of race as something that inflects and diversifies processes of gentrification in a subsequent paper (SLATER 2009), in which he praises in particular the work of Michelle BOYD (2005). Her ethnography of a gentrifying African-American neighborhood reveals the intimate connections and interpenetrations of race and class. SLATER (2009, 301) argues that BOYD "rejects as an *illusion* the contention that gentrification is happening in the interests of – and with the approval of – the poor black residents it threatens to displace."

This empirical question, of whether African-Americans seek gentrification as a re-investment in their neighborhoods, or whether they fear it as a potential displacement, has haunted debates over gentrification in the United States for a long time because of its possibility to reduce the debate of gentrification to one of purchasing power and markets (which in the U.S. is rarely a debate at all) by abstracting out the historical legacy of African-American oppression (which can have great political resonance). In the following section this paper shifts its focus to a representation of this question from popular culture, which is one of the ways in which African-American consent to gentrification has been produced as commonsense.

THE 'RELEVANCE' MOVEMENT AND *MADNESS IN THE SLUMS!*

The 1970s saw the emergence of a comic book industry that aspired to less fantasy and more social realism (LOPES 2009). The so-called 'relevance movement' saw the more overt introduction of contemporary social issues into the pages of comics such as *Captain America*, which had previously been a wartime comic in which the star-spangled hero battled against Nazis, Japanese, and Communists (DITTMER 2007a, 2007b) or alternately a Buck Rogers-like 'man out of time', commenting on the turbulent 1960s from the perspective of a patriotic American who was raised in the Depression and fought in the 'Good War' (DITTMER 2009). Race played an important part in the relevance movement, and *Captain America* is credited with introducing the first African-American superhero in comics, the Falcon.

> Marvel [Comics] wanted to represent African American characters in their stories in a realistic manner. The Falcon was not the fantastic king of an African nation, as was the Black Panther; he was not a character that could be looked up to (MCWILLIAMS 2009, 70).

The Falcon served as erstwhile partner for Captain America (the term 'sidekick' is definitively eschewed by the writers) for 105 issues, tying the blonde-haired, blue-eyed nationalist hero, Captain America, to specific problems in Harlem (the domain which the Falcon saw as his particular turf; his alter ego, Sam Wilson, was a social worker). Through this grounding in an African-American ghetto, stories in *Captain America* were able to address racial inequality and the nature of power.

> In nearly every early issue of [the Falcon's] run, there is some reference to Harlem, or scenes in which the Falcon mentions his 'black brothers'. The other side of the argument is that these stories walked a delicate balance. They may have contained blaxploitation, but they also exposed a reality to an audience that would have no other exposure to these types of social problems (MCWILLIAMS 2009, 74)

One of those social problems taken up specifically was that of displacement and gentrification; however as MCWILLIAMS indicates this narrative stands on a knife edge between raising social issues and legitimating 'white' authority as well as displacement and gentrification.

With a cover date of January 1971, but in practice published several months prior (and thus in late 1970), *Captain America #133* contained a story entitled 'Madness in the Slums!'. The story begins with the supervillain MODOK explicating his desire to make Captain America suffer. MODOK is essentially a huge swollen head contained by a hover-chair which he uses to move around. As his misshapen physiognomy indicates, MODOK's key attribute is his brain (hence his name, which stands for 'Mental Organism, Designed Only for Killing'), a common binary opposition to the physicality of the superhero in these genre narratives. MODOK's new plan to punish Captain America is to exploit the racial divisions within the United States in order to ruin Captain America's reputation. To do so, MODOK creates a three-story tall clay monster and animates the behemoth with the mental power at his command. He names his creation 'Bulldozer, the readers' first indication of how this story might develop in ways relevant to displacement and gentrification.

Bulldozer is dropped off by MODOK's criminal organization, AIM (Advanced Idea Mechanics), in New Jersey, from whence Bulldozer makes his way to the African-American neighborhood of Harlem, home to Captain America's partner, the Falcon. There, Bulldozer begins destroying housing, while vocalizing pre-programmed phrases such as 'power to the people!' and 'AIM must befriend the oppressed!' (LEE and COLAN 1971a, 14). Captain America and the Falcon rush to intervene, as one of the definitions of the superhero is that he or she works to protect property, political orders, and the status quo (WOLF-MEYER 2003; DITTMER 2007c). Complicating matters for Captain America and the Falcon is the fact that the African-American residents of the neighborhood interpret Bulldozer's platitudes as genuine, and see his actions as being altruistic (Figure 1). Captain America and the Falcon are heckled by Harlem residents when they try to stop Bulldozer, as are the police when they arrive. Back at his headquarters, MODOK celebrates the way he has positioned Captain America as a reactionary force by aligning Bulldozer with the revolutionary potential of the frustrated slum-dwellers, even as he acknowledges his actions will only lower their quality of life (Figure 2).

Figure 1: Bulldozer destroys the housing of Harlem and is understood as a hero by the residents (LEE and COLAN 1971a, 15). Copyright 2013 Marvel Characters, Inc. Used with permission.

Figure 2: MODOK celebrates the way he has positioned Captain America as a reactionary force by aligning himself with the residents of the ghetto (LEE and COLAN 1971a, 16). Copyright 2013 Marvel Characters, Inc. Used with permission.

Captain America realizes that further direct confrontation between the authorities and Bulldozer will only exacerbate racial tensions, and so he calls on the police to stop fighting Bulldozer. Then, in a horrific *deus ex machina* plot device, Captain America's fellow superhero Iron Man devises an energy detector that allows Captain America to trace the mental energy that MODOK uses to animate Bulldozer. He uses this to uncover MODOK's location and defeat him, thus ending Bulldozer's 'urban renewal' project.

Neither the rapid (and unsatisfying) conclusion of the conflict nor the narrative's denouement address the social issue provocatively raised by the storyline, the support of the African-American community for the destruction of the 'slums' in which they lived. Brian CREMINS (2002, 243) emphasizes the tensions found in this very issue of *Captain America*:

> What results is a sometimes confusing, bizarre reading experience. […] In the midst of all this action, does the reader have the time or the inclination to reflect on the serious issues [writer and editor Stan] Lee presents? Is it possible for a medium designed as a means of childhood escapism to incorporate adult concerns and complexities without losing its entertainment value and appeal?

However, one letter-to-the-editor three months later raised the question of racism in the portrayal of the Harlem residents:

The Bulldozer is destroying Harlem, and the citizens are cheering him on and turning on Cap and [the Falcon]. Think about that for a moment. A three-story giant comes into town, destroying your home, killing your family and friends. No matter how little these seem to be, no matter how oppressed the people are, they would not give up their only possessions and their loved ones to a stranger. Do you honestly think that black people are really that blinded by hate and fear to actually back up a foreign, ugly, super-powered maniac who comes from nowhere, smashing, destroying, and killing? The giant utters a few riot slogans in a mechanical, unfeeling, unemotional voice, and gains the trust and admiration of an entire community, who are noted for their distrust of an outsider. Highly questionable! (Frank Nocerno, in LEE and COLAN 1971b, 21)

This accusation was rebutted by writer and editor Stan LEE, who articulated a more complete explanation for the behavior of the Harlem residents than was made in the story itself:

We suspect, Frank, that if you think this story is totally unbelievable, you've never taken a walk through Harlem. We have. The 'homes' Modok was destroying – the unheated, unplastered, unsafe, unsanitary homes, many lacking hot or even running water – caused relief, in a sense, to tenants who live at the mercy of uncaring and apathetic landlords. The Bulldozer was not killing family and friends as you inferred from our story – that would have made the people's support of him illogical.

In the eyes of the ghetto, Captain America represented the establishment – that same establishment which includes irresponsible landlords. And so, the emotional connection between the two caused the community to back any power opposed to the continuance of their plight. Do not sight the desperate action of desperate men as unbelievable. Sometimes it is the only way up (LEE and COLAN 1971b, 21).

The two positions articulated above are in some ways speaking past each other; the letter writer is arguing that as much as the Harlem residents might resent their living conditions they would prefer them to being forcibly displaced by a giant monster. Drawing on the logical inference that the residents' possessions were also being destroyed, the letter-writer implies that white renters would not be represented as behaving in this way, and thus the story is unrealistic and possibly racist. Stan LEE's response emphasizes the role of corrupt landlords who extract rent from Harlem without re-investing, as well as the political context that enables the extraction of wealth from one place and its accrual in another. However, to justify this understanding of the housing system in Harlem LEE has to make the rather implausible claim that Bulldozer's actions were not dangerous to anyone in the area.

CONCLUSIONS

Setting aside the question of danger to Harlem residents in 'Madness in the Slums!', it is worth considering the relationship between race and displacement in the narrative. This relationship is contextualized by the spatio-temporality of the story's conflict: Harlem in the early 1970s. In an era of red-lining and residential segregation (MASSEY and DENTON 1994), in LEE's argumentation it is seen as axiomatic that if landlords build new residences then they will be occupied by the same people who have been displaced by Bulldozer, and not by in-migrants drawn by the possibility of gentrification. Of course, it is also entirely possible that middle-class African-Americans could move into the new houses, as gentrification can occur within racially homogenous areas as easily as between racially diverse ones.

Of course, it is perhaps too much to expect a sophisticated analysis of gentrification and its dynamics to appear in a superhero comic book, but nevertheless it is to 'social relevance' that these comics aspired. 'Madness in the Slums' could leave the reader with the impression that displacement and reinvestment are a good, and even necessary, response to conditions in the African-American ghetto. Brian CREMINS (2002, 245) argues that the relationship between the text, the presumed reader, and the ghetto can be summed up as one of catharsis:

> [T]he racialized performances in a novel like *Huckleberry Finn* or an issue of *Captain America* cleanse the anxieties and fears at the heart of white society, or at least the stirrings of guilt in a white, middle-class male reader first growing aware of a world outside the four-color walls of the Marvel Universe.

In this way, 'Madness in the Slums' can work to normalize white power even while purporting to do the opposite.

Perhaps it is worth considering a version of the question that SLATER (2009, 301) has asked of those who argue that gentrification brings investment in the form of enhanced city services – "why does it have to be gentrification that brings better services?" Similarly, why did LEE (and artist Gene COLAN) sculpt the politico-moral pivot of the narrative around displacement, when he could have had Captain America and the Falcon take on the corrupt landlords directly, compelling them to reinvest in their properties without displacing the current residents? This poverty of the narrative imagination parallels the poverty of the political imagination when it comes to the right to stay put; implementing such a right will require openness to new ways of thinking about property and residence, as well as attention to the ways race inflects both representations of gentrification and the ways such a liberal, universal right would be implemented fairly.

REFERENCES

BOYD, M. (2005): The Downside of Racial Uplift: meaning of gentrification in an African American neighborhood. *City & Society* 17 (2), 265–288.

CREMINS, B. (2002): "Why have you allowed me to see you without your mask?" *Captain America* #133 ang the Great American (Protest) Novel. *International Journal of Comic Art* 4 (1), 239–247.

DITTMER, J. (2007a): "America is safe while its boys and girls believe in its creeds!" Captain America and American identity prior to World War 2. *Environment and Planning D, Society & Space* 25 (3), 401–423.

DITTMER, J. (2007b): Retconning America: Captain America in the Wake of WWII and the
    McCarthy Hearings. WANDTKE, T. (Ed.): *The Amazing Transforming Superhero! Essays on
    the Revision of Characters in Comic Books, Film and Television*. Jefferson, North Carolina
    and London, 33–51.

DITTMER, J. (2007c): The Tyranny of the Serial: Popular Geopolitics, the Nation, and Comic Book
    Discourse. *Antipode* 39 (2), 247–268.

DITTMER, J. (2009): Fighting for Home: Masculinity and the Constitution of the Domestic in the
    Pages of *Tales of Suspense* and *Captain America*. DETORA, L. (Ed.): *Heroes of Film, Comics
    and American Culture: Essays on Real and Fictional Defenders of Home*. Jefferson, North
    Carolina and London, 96–115.

GRAY, N. (2009, 18 November): Going Nowhere or Staying Put? *Mute Magazine*, from
    http://www.metamute.org/en/content/going_nowhere_or_staying_put.

HARTMAN, C. (1984): The Right to Stay Put. GEISLER, C. and F. POPPER (Eds.): *Land Reform,
    American Style*. Totowa, New Jersey, 302–318.

HARTMAN, C. (2003): The Case for a Right to Housing in the United States. LECKIE, S. (Ed.):
    National Perspectives on Housing Rights. Leiden, Netherlands, 141–162.

IMBROSCIO, D. L. (2004): Can We Grant a Right to Place? *Politics & Society* 32 (4), 575–609.

LEE, S. and G. COLAN (1971a): Madness in the Slums! LEE, S. (Ed.): *Captain America #133*. New
    York, 1–22.

LEE, S. and G. COLAN (1971b): To Stalk the Spider-man. LEE, S. (Ed.): *Captain America #137*.
    New York, 1–22.

LOPES, P. (2009): *Demanding Respect: The evolution of the American comic book*. Philadelphia.

MANZO, L., R. KLEIT and D. COUCH (2008): "Moving Three Times Is Like Having Your House on
    Fire Once" The Experience of Place and Impending Displacement among Public Housing
    Residents. *Urban Studies* 45 (9), 1855–1878.

MASSEY, D. and N. DENTON (1994): *American Apartheid: Segregation and the making of the
    underclass*. Cambridge, Massachusetts.

MCWILLIAMS, O. (2009): Not Just Another Racist Honkey: A history of racial representation in
    Captain America and related publications. WEINER, R. (Ed.): *Captain America and the
    Struggle of the Superhero: Critical Essays*. Jefferson, North Carolina and London, 66–78.

SLATER, T. (2008): 'A Literal Necessity to be Re-Placed': A Rejoinder to the Gentrification
    Debate. *International Journal of Urban and Regional Research* 32 (1), 212–223.

SLATER, T. (2009): Missing Marcuse: On gentrification and displacement *City* 13 (2–3), 292–311.

WOLF-MEYER, M. (2003): The World Ozymandias Made: Utopias in the Superhero Comic,
    Subculture, and the Conservation of Difference. *Journal of Popular Culture* 36 (3), 497–517.

Jim Craine / Giorgio Hadi Curti / Stuart C. Aitken

COSMOPOLITAN SEX, MONSTROUS VIOLENCE AND NETWORKS OF BLOOD

"We are entering savage times," quips a video technician at the beginning of Canadian director David CRONENBERG's *Videodrome* (1983) as he switches on bootlegged snuff-TV images for television producer Max Renn (James Woods). What ensues is a disturbing urbanity of technology, pathology, religion, lust, death, desire, addiction, murder, sexuality, sadism, cancer, masochism and conspiracy that, sixteen years before *The Matrix* (1999), speaks to the transformation of the corporeal city through a seemingly pure and, without a doubt, more monstrous electronic and sexual energy.

Anticipating the virtual worlds of William GIBSON and Donna HARAWAY's human/technology hybridity, CRONENBERG explores the monstrous melding of bodies, technology, sex, death, interior, exterior, and the virtual. With CRONENBERG, we explore the contours of two kinds of monstrosity (a double articulation of sorts) and suggest that it is about the savagery through the construction of eyes/I's that are elsewhere in *Videodrome* that serve as a "spectacular optics" to global power as well as intimate "windows on the soul" of the city. The spectacular, monstrous, unblinking eye: BAUDRILLARD's (1983) "Hell of the Same," Kaja SILVERMAN's (1981) "decentered (in)human subjectivity," Guy DEBORD's (1983) "science of domination." While the monstrous eye/I is as much something we seek, create and consume as that which seeks, creates and consumes us, CRONENBERG's work is a poignant indictment of the forces that turn inside-out as corporate mayhem ensues in a cautionary tale of the power that resides within the psyche, and indeed, the lived city. Like a Deleuzian dream of Bergsonian virtuality, CRONENBERG's solution is a double articulation that has capacities to be destructive and generative, both illness and remedy. Ostensibly about the power of images and video representations, *Videodrome* deforms into a monstrous parody of the co-creation of technology, murder, flesh and desire that goes well beyond representations to strike at the affective heart of our dark and light urban places. Through this, it communicates perhaps more clearly than any other media expression – and, in the process, reveals its indispensability in understanding – the immanent Janus-faced capacities of emerging digital technologies of visuality we find ourselves and our cities entangled in today.

In its own particular way, *Videodrome* provides a carto-cinematic multidimensional cartography, a set of landscapes that present information and sensations for multiple social, cultural and urban phenomena – a data set that can be explored, analyzed and interpreted through registering effects on the human processing system. To put it another way, *Videodrome* is a space of the fantastic, where traditional notions of distance no longer hold the same effect and the imaginary and the speculative become visible and tactile through forces of images and affects that play out on, in, between and through viewer bodies and viewing screens. Approaching *Videodrome* through such understandings allows us investigation into the repulsive subterranean phenomenon of its world's snuff-TV (the show Videodrome within the film *Videodrome*), while simultaneously working as a means of raising questions about our own urban fascinations with a particular form of the monstrous: sex and violence. It is in a double articulation that this monstrous envelops and enfolds *Videodrome*: in one sense, it is a violently sexualized monstrous of deception, manipulation, exploitation, and passivity; in another, a wonderfully (pro)creative monstrous of violent affectivity, difference, surprise, and activity – in both, *Videodrome* is as much a commentary on the double monstrosity of the city as it is the dual powers of media that continually challenge or reinforce the urban monstrosities of our own creations.

A MONSTROUS DOUBLE

What is it about Videodrome that makes it so monstrous in the first sense? It is precisely, as feminist soft-core pornographer Masha (Lynne Gorman) tells ostensible protagonist Max Renn, that "it has a philosophy." It is a dangerous and violent philosophy not of the Greeks – it is not φιλοσοφία [*philosophia*], or the love of wisdom. Rather, it is a nefarious philosophy, a malicious, insidious, grotesque philosophy masquerading as moral wisdom built on and through ties of paranoia, anger, and hate instead of love. It is philosophy as ideology: an ultimately hollow wisdom of passive closure and isolation which seeks to detach mental and bodily capacities from powers of activity and creativity through a very literal sadism: rationalism as "an internal necessity that…evolves the idea of a delusion" (DELEUZE 1989a, 27); it is a calculating philosophy of resentful resolution, hegemony, and hierarchy; it is the philosophy of the tyrant, the murderer, the rapist, and the pedophile. Indeed, "The subject of [*Videodrome*] is hardly human action: it is instead…the structures of external power and control to which the individual (in body *and* soul) is subjected" (BUKATMAN 1990, 202). It is precisely through the subject(s) of *Videodrome* (that is, through *Videodrome* as filmic narrative and Videodrome's control over and effects on Max Renn) that this first monstrous unfolds.

But the *subject(s)* of *Videodrome* must be distinguished (though not detached) from its *capacities* to affect. Understandings of cities over the last one hundred years or so have become a focus for coming to terms with quite a number of different things about ourselves, including our dreams, fantasies and psychoses. The important point about Max's addiction to the spectacle of *Videodrome* is that it involves choice but it also stimulates adrenaline, dopamine and all the other neurotransmitters that propel us to come back for more. *Videodrome* presages our cosmopolitan intrigue with spectacle and the commodification of desire, and it highlights the dystopian underbelly of an enlightened cosmopolitanism through addiction and horror, technology and corporeality, sex and death. But in a deeply ironic way it is also about hope and redemption, about moving forward through and with urban decay and dystopia. In this sense, it is important to come to terms with the idea of Gilles DELEUZE's notion of a double articulation, when the content (e.g. an institutional frame, perhaps part of the heart of global capitalism) is overpowered and expression (feelings and desires) overpowers (MASSUMI 1992, 152).

Like Max Renn, we as urban creatures in many ways have become physically addicted to the monstrous in its first sense – and we, like Renn, become fascinated with Videodrome's exploitative monstrous: a sadomasochistic television series apparently originating from 'exotic' Malaysia (but in actuality is coming from Pittsburgh, Pennsylvania) that exhibits a plotless single-camera study of the infliction of torture on human beings. We, like Renn, are mesmerized by Videodrome – its voyeuristic appeal, its compelling realism and obviously low production cost – we *want* it, we seek out its illusive imagery, sounds and performances. Ultimately, however, it dawns on us – like Renn – that Videodrome is no illusion: it is authentic snuff-TV – a realization that becomes more chilling as we watch its violent whippings, beatings, strangulations and genital electrocutions; an unfolding of a negative, scientific philosophy which ultimately supports DELEUZE's (1989a, 124) notion that the psychoanalytic sado-masochistic pairing is plainly artificial, and that "[t]*he sadist has no other ego than that of his victims*; he is thus monstrously reduced to a pure superego which exercises its cruelty to the fullest extent." Despite these sadistic and monstrous materializations – and even with our discomfort – we smirk as Renn exclaims with prescient candor and clarity: "it's all just torture and murder. No plot, no characters. Very, very realistic. I think that's what's next." It is precisely this realization and *Videodrome's* subsequent stimulation of a doubly monstrous urban landscape that present sites to be explored for cultural and material relevance and meaning.

WE LIVE IN OVERSTIMULATED TIMES

Describing *Videodrome* without watching the film runs a risk of creating some Monty Pythonesque world of humor and debauchery (exploding entrails, oozing heads, vaginal openings in stomachs, sex-starved televisions). This is not a comedy, nor even a black comedy. The film is the epitome of the monstrous, or rather a double monstrous, using images to move well beyond the suspenseful shock and horror of Alfred Hitchcock or George Romero. For CRONENBERG, the monstrous is never comfortably external: it is not one of Danny Boyle's raging zombies, nor can it be tacked down to one of John Carpenter's layers underneath the city or beneath the ice. It is not an invasive alien species or a predator from another planet. Nor is the monstrous totally from within, the completely pathological or quintessential embodiment of evil. His monster resides in the spaces *between* us, our culture, our society, our cities – our over-stimulations and relations of want and desire.

As CRONENBERG's movie progresses, Max is pulled further and further into a seemingly hallucinatory world where the virtual holds sway. Part virtual in the Gibsonian sense of technology induced virtual reality; it is also virtual in the Deleuzian sense of a pure immanence that is always in the-making. The reality TV Videodrome snuff show is quickly superseded by layers of intrigue and debauchery, sex and death, guised in part as a redemptive religious moment emanating from the underside of Toronto's city streets through the Cathode Ray Mission, but more likely pivoting around corporate control of people's desires. *Videodrome*, then, is also about the moment when, through spectacle, commodity is on the cusp of attaining DEBORD's (1983, 42) *total* occupation of urban life, where "Modern economic production extends its dictatorship both extensively and intensively." If LEFEBVRE (1996) sees the right to the city as the right to desire (amongst other things) then *Videodrome* bastardizes that right by moving desire into a realm of addiction, debauchery and control that is guided and controlled by global corporate capitalism.

The heart of DEBORD's global corporate capitalism is about control of spectacle and, ultimately, the control over desire. It thus confronts a larger cosmopolitan conundrum that arises from the *supermodernity* that is the heart of global capitalism (HUBBARD 2006, 168), one which revolves around how to move forward and when to stay put: questions which have both extensive and intensive implications. *Videodrome* precisely plays out these internal and external implications through a kind of cosmopolitan conspiracy that resonates with Julia KRISTEVA's *Powers of Horror* (1982) in the sense that it examines the choices we make around our desire for horror, castration, marginalization and exile. For KRISTEVA, being forced to face abject horror such as a corpse that we recognize as human is an inherently traumatic but also tantalizing experience. The abject is situated outside of the symbolic order of things to which we are simultaneously drawn and repelled; the horror is external but so too is it part of our own intensive embodiments. *Videodrome* takes tantalization to its most horrific extremes and, in so doing, shows us the heart of corporate cosmopolitanism.

Spectacular Optics

The lips of Nicki Brand (Debbie Harry) mouth the construct "we live in over-stimulated times" that presages the addictive, body-penetrating, technology melding, deviant sexualizing, delusional and eventually murderous viral romp upon which Max Renn is about to embark. Concerned by a new satellite TV competitor to his mildly pornographic Civic TV[1], Max seeks to learn more about the origins of the competing signal. Fascinated by its depiction of graphic sadism, masochism and murder, Max thinks that maybe Videodrome has an edge that Civic TV lacks. He discovers the Cathode Ray Mission, and the deceased Dr. Brian O'Blivion (played by Jack Creley and modeled after Marshal McLuhan) who talks to Max through a library of videotapes about the larger conspiracy behind what he calls "Videodrome Syndrome."

Although deceased, videotapes enable O'Blivion to return "back into the world's mixing board." Mirroring McLuhan's sentiments, he declares that "the television screen has become the retina of the mind's eye" and suggests that humankind's next evolutionary stage is through biology and technology and on to pure electronic energy. The homeless denizens of the Cathode Ray Mission follow the teachings of Dr. O'Blivion and from televisual booths the urban masses receive their stimulation.[2]

Nicki, taken in and consumed by Videodrome, beckons to Max through his pulsating, voluptuous television screen.

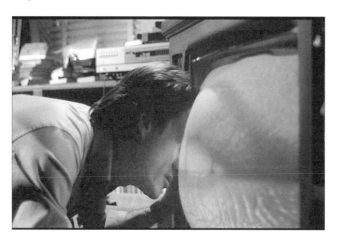

This is the beginning of his somatic hallucinations. Apparently, Videodrome Syndrome is realized through drug-like psychotic illusions perpetrated through

---

1   Cronenberg modeled Civic TV after Toronto's CITY-TV, a television station that began in 1972, owned and operated by Rogers Media. CITY-TV was best known for its unconventional approach to TV news and programming, an approach that continues today and carries over to other stations in the local CityTV system. Rogers Media is now one the largest media corporation in Canada, particularly in terms of wireless and cable television. CITY-TV was the first network in the world to switch to cable. Like CITY-TV, Cronenberg's Civic TV is intensely local with global aspirations.

2   While this is not Videodrome – what is presented to us with the Cathode Ray Mission speaks just as loudly to the other side of debauchery and sex: like Videodrome the tactics of the Cathode Ray Mission are about televisual overstimulation and conquering death by embracing a new kind of urban embodiment.

watching Videodrome's indistinct, disturbing, and yet tantalizing, sadistic and masochistic images. Now less concerned about the plight of his TV station given the prospects of Nicki's murder on the Videodrome stage, Max tracks down its headquarters in an ophthalmology and eyewear store called *Spectacular Optics* where promoter and TV personality, Barry Convex (Les Carlson), tells him "Love comes in at the eye, and the eye is the window of the soul." There is a sense of something that is simultaneously wholesome and perturbing in the proclamations of this contemporary exegetes.

Max's hallucinations continue. In a disturbingly graphic Freudian moment, his gun is consumed by an oozing vaginal-opening in his stomach to return later as a phallic, cancerous techno-appendage at the end of his arm. CRONENBERG's work is about creating a monstrous imaginary through compellingly repulsive graphics that pre-empt the flat computer-generated images of today's horror genres.

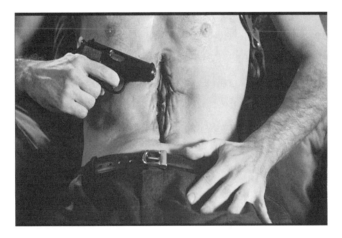

CRONENBERG's narratives and images thus work at a subtle cerebral and visceral level to confound and disturb our hold of reality. By looking at what *Videodrome* pre-sages, we see it as an icon pointer to the monstrous simultaneously as an intensive bodily ordinate intersected by urban and media over-stimulation and creativity. It moves horror and the monstrous away from external threats (communism and alien invasions of 1950s movies, conspiracy theories of the 1960s, or unstoppable zombies from the 1970s) to a pathology that is uncomfortably situated both internally and externally. It is the form of horror articulated in hyper-masculine violence that is simultaneously shocking and disturbingly comfortable (AITKEN 2006).

THE MONSTROUS IMAGE: LANDSCAPES OF EXPLOITATION AND CONTROL

By considering how *Videodrome's* monstrous bodies and landscapes are created, how desires, subversions and violence are inscribed in bodies and the city – metaphorically and literally – and how they are then used to perpetuate a reality that is accepted as 'natural' by the populations of those landscapes – in other words, how they create *us* – we can determine what urban landscapes mean and how these meanings and images are geographically visualized and used to shape culture. Pointing to this, MITCHELL (2000) indicates that film is a space in which values are produced and contested – cinematic space can be approached as ideological battlegrounds, or insidious monstrous spaces enhancing the ability to organize, control and shape social and urban reproduction; and we, like Renn, are concerned with who controls, has access to, and definition over our spaces of relation and communication.

This becomes central as these spaces have increasingly moved *inside* – inside the movie theater, inside the home theater, inside the digital structure of the hard drive, DVD and Blu-ray – and are thus amenable to monopolized control over production, affects and meaning. Indeed, confronting the capacities of control or creativity of these new media forms and their organizing and divisive spatialities is becoming more and more vital with their ubiquity. The relevance of this is obvious when recent comments by Intel Corporation's chief technology officer Justin RATTNER are considered:

> TV will remain at the centre of our lives and you will be able to watch what you want where you want…We are talking about more than one TV-capable device for every man and woman on the planet…People are going to feel connected to the screen in ways they haven't in the past (SHIELS 2009).

These new spaces and complexes through which (tele)visual data are captured, processed, manipulated, and viewed and by which meanings are disseminated and transferred have implications for the ways we know and apprehend the city. As MITCHELL (2000, 86) stresses:

> If the means for organizing an audience, either in favor of or against some set of social meanings, either as a passive audience for mass-produced entertainment, or as an active force for social change, themselves exist primarily for the production of profit, then the *types* of entertainment, possible resistances, and so on, are highly *biased.*

The study of landscapes as texts has as its central site of inquiry a distinctively static scene – that is to say we observe the landscape as fixed space. In static landscapes, the landscape passively 'represents' some history or another. Giuliana BRUNO (2002, 173) argues that the realism of 19TH century landscape painting says more about the 19TH century than it does about landscapes. Landscape is constantly in motion through geography and the body, she points out, because they share a tangible territory of lived space that is traversable and thus mobile. In reality, then, landscape itself is an active and dynamic agent in constituting history (CURTI 2008a, 2008b), while also serving as a symbol for the needs and desires of the people who live in it (or who otherwise have a stake in transforming or maintaining it). *Videodrome* thus provides us with a very significant tool with which to engage cinematic landscape biases, one that instills an extremely critical shift into the geography/film metaphors of landscape representation.

Landscape, as MITCHELL (2000, 93–94) explains, is best seen as

> both *a work* (it is the product of human labor and thus encapsulates the dreams, desires and all the injustices of the people and social systems that make it) and as something that *does work* (it acts as a social agent in the further development of a place).

BRUNO (2002, 81) makes the point that with film the female (and male) subject are increasingly able to map herself (and himself) into "the epistemology and erotics of mobility." And Videodrome is indeed a social agent at work and doing work upon the evolving and mobile geotechnological and erotic landscape that, as HARAWAY (1997, 4) indicates, "require[es] a broad technoscientific research and communication apparatus." Our futures are inexorably linked to the futures of image production and televisuality, and are thus tied to the advances of science in a very unique manner. As the subject(s) of *Videodrome* (and Videodrome) reveal, there exists a very real threat to our spaces of cultural production from monstrous forces of closed rationalities and moralities. As video technician and Videodrome co-conspirator Harlan (Peter Dvorsky) relates to Max:

> North America's getting soft, *patron*, and the rest of the world is getting tough. Very,very tough. We're entering savage new times, and we're going to have to be pure and direct and strong, if we're going to survive them. Now, you and this cesspool you call a television station and your people who wallow around in it, your viewers who watch you do it, they're rotting us away from the inside. We intend to stop that rot.

It is precisely through the television, through the televisual signals and images of the show Videodrome, that a hegemonic monstrous of government and corporate conspiracy does work in the city to not simply control, but to ontologically negate – to exploit and murder – those deemed social and ideological cancers through an efficient deviousness of monstrous cancerous tumors, themselves the effect of viewing Videodrome. It is a particularly sadistic virus which envelopes Videodrome as its superego and id come together to monstrously feed its meaning and purpose through the control and negation of its victims' egos (read difference):

> The viral metaphor is strikingly apt when applied to *Videodrome* – the literalized invasion of the body by image, and the production of tumors which produce images. Image is virus; virus virulently replicates itself; the subject is finished. We remain trapped within a universe which seems to be somebody else's insides (BUKATMAN 1990, 207).

In this way, *Videodrome* fantastically underscores MITCHELL's arguments surrounding entertainment and control and critically points to "Television's main use…for social control rather than artistic invention" (BOGUE 2003, 196).

Thus, we are confronted with interactive media in a very perverse way as the TV evolves from a passive sex aid to an active sex partner; there is the constant collision between flesh and technology (through pain, desire, and lust). It is perhaps more metaphysical here, but CRONENBERG revisits this theme many years latter in his adaptation of J. G. Ballard's *Crash*, while *Videodrome* further references 50s *film noir* via *Kiss Me Deadly* in that the tumor-inducing Videodrome signal recalls the "Pandora's Box" that ignites that film's plot. Moreover, in many ways Videodrome now seems like the horrible culmination of what we now term "Reality TV," where strangers (i.e. contestants) are put in humiliating situations and sometimes find themselves figuratively 'killed' (i.e. Donald Trump's "You're fired' or the judges of American Idol crushing the dreams and confidences of another 'little person' under the gaze of millions of TV-bound voyeurs). DELEUZE (1995, 179) pointed out many years ago, "If the stupidest game shows are so successful, it's because they're a perfect reflection of the way businesses are run." Reality TV and game shows come to naturalize competition and imbue life with a sick joke: the logic of the capitalist spectacle has become a way of life, and the city its most intensive ordinate. Indeed, it is not a big jump to see how this representational violence could easily become concrete – the photos from Abu Ghraib have a disturbing similarity to what is staged in the Videodrome show (particularly the one of the hooded prisoner with his genitals wired for electric shock).

The film also touches on our dependence on technology: the gun that attaches itself to the James Woods character certainly brings to mind a generation of young people flowing through cities manically attached to their cell phones and iPods and video games, and it also taps into our memories of that parental warning "Don't sit too close to the TV" – which thus plays into the theme of the image – and its effects – as deviously monstrous. In this way, *Videodrome* certainly plays into an engagement with a potentially harmful becoming-virtual of urban culture; and, indeed, the transformative process could be seen as becoming monstrous a la Frankenstein; Renn literally becoming his virtual counterpart where "[b]ody and image become one, a dissolution of the real and representational" (BUKATMAN 1990, 207): the ideal outcome of a media theory that bypasses activity and affectivity for a realm of reducing media to representation and a particular sense of a closed, passive, harmful, and hegemonic monstrous.

THE MONSTROUS IMAGE: LANDSCAPES OF AFFECTIVITY AND CREATIVITY

But *Videodrome* also offers a distinct challenge to this hegemonic monstrous urban life of the image, media, television and televisuality through the very real physical landscape of a creative monstrous impelled through our relationships with sexuality, desire, violence and technology, and which include televisual spaces from the home to the workplace to places of leisure. The film portrays a landscape of rapid advances in the way we visualize reality: Max Renn's need for remotely sensed imagery mirrors our own – evolving from the videographic into the digital – thus creating new models and new principles related to how we visualize reality and its relationships to the mind and body. Indeed, as in many CRONENBERG films, Renn's bodily transformation from Videodrome and in *Videodrome* "cannot be completely subsumed within the mind/body dualism of Cartesianism" (BUKATMAN 1990, 202). And it is precisely this ontological movement away from Cartesian dualisms and understandings of bodies and space that point to a new (becoming) vitality of a creative monstrous in relationship with new media and televisual urban spaces.

Our new forms of digital film such as the DVD and Blu-ray give us a mathematical algorithm simulating or modeling the geometrical form of the image it generates thus presenting itself as a new source of knowledge. Thus, there is a new power associated with the image and these new technologies; like those found in *Videodrome*, they may actually blur or even invert the primal relationship of the image and the city:

> As *Videodrome* so graphically illustrates, 'simulation' is not simply an additional layer of special effects that is overlaid upon the body of the earth, like a sediment or mask; it turns the inside out and the outside in. This invagination is neither inversion nor reversal, but Möbius spiralling: through a simple twist of fate, the ins(l)ide gives way to the outs(l)ide – and vice versa (DOEL 1999, 95–96).

While representationally a hegemonic monstrous plays out in *Videodrome* through a "Möbius spiralling" of televisual images which take a devious role in (re)shaping characters, bodies, and plots, becoming the defining force providing the context for the social, political, and cultural urban exchanges that define and (re)create its spaces and societies, this relationship can not and need not be reduced to this single sense of the monstrous. As Ronald BOGUE (2003, 196) explains in his reading of DELEUZE and cinema, "Each medium has its potential for creative development," and "much of the potential of television [has been so far left] unexploited."

Videodrome "works" and "does work" because it is technogenic, it underscores the co-evolutionary relationship between humans and the technologies that shape urban culture and material realities – it moves affectivity into the digital realm and therefore beyond our traditional notions of distance or our previous understandings of the relationships between technology, sex, and violence. *Videodrome* is significant in relation to DELEUZE's (1989b; 1995) conception of the affection-image, and the entire sublimation of affection into the image that lies at the heart of DELEUZE's effort to expand the scope and philosophical vocation of cinema. Indeed, *Videodrome's* liberation of affectivity as the condition for the emergence of perception and its creative monstrous imagery uncovers the relationship through which our confrontation with the violent image can be transformed into an intense and vitalizing affective experience. *Videodrome's* liberation of affectivity restores continuity to the experience of time. In exposing the time-image's dependence on an act of virtualized affectivity by the viewer, *Videodrome* reveals embodied affectivity to be the condition of possibility for the apprehension of the time-image. Yet, it is vital to note that this embodied affectivity of a different sensibility of time is no end-point, it is not the affective body of phenomenology:

> The phenomenological hypothesis is…insufficient because it merely invokes the lived body. But the lived body is still a paltry thing in comparison with a more profound and almost unlivable Power *[Puissance]* (DELEUZE 2003, 39).

This configuration of the time-image, an important aspect of any carto-cinematic presentation, leads directly back to Bergson's fundamental conceptions of the body as the union of perception with affection, and hints at powers beyond the lived body. Videodrome (the television show) consists of such affective powers and it significantly expands Bergson's crucial insights into the bodily basis of intelligence – including technical intelligence – revealing it to be even more central as we more actively engage digital data spaces in the multiple places of today's imagistic cities – from the home to the streets to restaurants, shopping malls and bars. Television then, like film, can be approached as "a modern cartography," because "its haptic way of site-seeing turns pictures [read images] into an architecture, transforming them into a geography of lived, and living, space" (BRUNO 2002, 8–9). In this way Videodrome also anticipates Doreen MASSEY's (2005) critique of Bergson's representation of space: it is not the quantitatively divisible, the representational, the static, but a living and dynamic force that is "part of it all, and itself constantly becoming" (MASSEY 2005, 28). *Video-drome* (and Videodrome) thus helps us appreciate the crucial importance of DELEUZE's study via MASSEY (and MASSEY's study via DELEUZE): the imperative to recognize the connection between human capacities and inhuman functions, including the inhuman and powerful functions of "space as the dimension of multiple trajectories, a simultaneity of stories-so-far. Space as the dimension of a multiplicity of durations" (MASSEY 2005, 24) – city/media space as symbiotically folding and unfolding organic and inorganic networks of blood.

DELEUZE's very Bergsonian analysis of cinema[3] shows that the material universe cannot simply be materialized via the image, and accordingly, that the focus must shift to the post-cinematic problem of framing *information* as an immanent part of "stories-so-far" in order to better understand the nature of embodied, affectively geovisualized *digital* images. *Videodrome* is thus expressive of both Bergson's theory of perception and embodiment and MASSEY's understanding of space and pushes us to accept new modalities of data that are created in digital and virtual forms, while also urging us to further explore apparatuses of engagement as the very catalyst for an empowering technical and sensuous transformation of the human through the affective consumption of new informational objects and spaces of the inhuman or the more-than-human (WHATMORE 2002).

What Videodrome *represents* is an image that has literally been plucked out of the aether – it ostensibly wasn't meant for our televisions to receive or for our eyes to see – but for certain the images are quite extreme, disturbing, and possibly even illegal – thus they are monstrous in many ways – yet we, in this case, as viewers of the film (*Videodrome's* subjects), or through Max Renn (Videodrome's subject) in the film itself, seek out these images – the basic concept is media's impact on our culture and technology's role in human beings' mental and biological evolution. Thus, the tie to individuation, cosmopolitanism and the neuro-aspect of geography *beyond [or before] representation* becomes central: can cities become/evolve into something monstrous – is their output monstrous because of how we create them, define them, limit them, and consume them? And perhaps the most important question: if the work the city does and which we consume is monstrous, *what kind of monstrous is it*?

The New Flesh

The issue of engaging our cosmopolitan sensibilities, of situating our urban angst, of fighting to stay put, of not evolving, of confronting the double monstrosities of our own digital and urban doings, of resisting the consumptive lure of Video-drome, is highlighted in the last few frames of the film.

---

3   See HANSEN (2003) for an in-depth reading of DELEUZE's cinematic [ab]use of Bergson.

Deluded, deranged and unable to save Nicki, Max kills several of his co-workers at *Civic TV.* At this point, we, like Max, are unable to untangle actuality from illusion, skullduggery from delusion. CRONENBERG creates an urban dystopia for the Max who has just realized what he has done. After his shooting rampage, we find Max wondering along a derelict railway track.

From the derelict rail track, he breaks into a deserted harbor4 and enters an abandoned tugboat. Here, he finds the discarded dwelling of some drifter, and he searches for a pack of cigarettes. This place feels like home to Max, a haven from his increasingly violent delusions.

---

4   This scene was filmed at the old industrial harbor of Hamilton, a few miles west of Toronto

Lifting some newspapers looking for a pack of cigarettes that still might contain a faggot, he discovers a TV remote. The look on his face tells us that although this place may be real, he is about to slip into delusion.

And, sure enough, not only is his old (voluptuous) television in the tugboat hold, but Nicki's image appears and she is telling him to embrace the new flesh. The image jumps to him standing, watching himself, but now he has the cancerous gun in his right hand. With Nicky's voice as a supplicant, the image puts the gun to its head and pulls the trigger.

The television explodes, strewing its guts5 all throughout the tugboat hold. The image cuts to Max, who slowly looks down at the cancerous gun/hand, which he raises to his head. The film ends with the pull of the trigger. Has Max evolved? Has he, like Nicky, become the new flesh? Is he now a disembodied series of signals? Is he resisting corporate cosmopolitan by joining the Cathode Ray Mission? Has he done society a favor by executing his co-workers at *Civic TV*? Has he found emancipation or was his act a suicide impelled and determined through the force of the spectacle, a slavery of a different form? These questions are quite trivial and not worth our attention, but in asking them we refocus attention to Max as a doubly monstrous embodiment of media and the city, and how they connect and intersect through an intimate vicariousness. CRONENBERG's genius is the way he leaves us on a horrific dangle through his use of Toronto and its environs as a trope around which the affects and effects of media and cosmopolitanism symbiotically congeal.

---

5   Actually, freshly butchered pig guts.

CONSIDERATIONS

In a very Kristevian sense, *Videodrome* engages the abject cosmopolitan by becoming a technological surrogate. We encounter radical and unsettling and traumatic challenges to our embodied understanding, our very capacity to make meaning out of the information of our technologies of visuality (e.g. Film, Television, but so too Google Earth, GIS, Global Position Systems). Because of the context of our encounter – our experience in contact with the digital, our affirming the power of the virtual in DELEUZE's sense, we become active. As we transform images into information (and then information into meaning) – creating a *supplementary* analogical connection to the digital – we become the image. This connection relates to an initial connection to a body that encounters imagery in its purely technical specificity: before any coherent information is exchanged, we undergo an immanent self-transformation that itself comprises the content of the creative monstrous as affectivity which is funneled into (while simultaneously transforming and extending) tangible and tactile information, which is to say, as information that becomes *meaningful* for the viewer. The meanings contained within the televisual spaces of *Videodrome* are virtualized through an affective process of technology and body – in representation as the body of Max Renn, and in a different, nonrepresentational way: *us*.

By reordering the attributes of space, we, through our proxy Max Renn, are presented with new and monstrous alternatives – new spatio-temporal technologies that emphasize both the relationships between technology, sex, urbanity and violence and the parallel attributes of body and mind. The film, and the television show within the film, disturb and confound our merely phenomenological experience by distorting and violating realism, calling attention to what works through reality both before and after reason. As filmic and televisual affections shock the haptic system and its geographic and cultural senses of space, they play a vital role in "shaping the texture of habitable space and, ultimately, mapping our ways of being in touch with the environment" (BRUNO 2002, 6). Thus, we can dismiss Videodrome co-creator Brian O'Blivion's rhetorical question, "After all, there is nothing real outside our perception of reality, is there?" as solipsism and replace it with questions of capacities, relations, affections and affects. With these questions we must not only reinterpret these new realities because they are constructive processes, but materially perform them through activity and creativity while emphasizing our awareness of a greater need to confront, encourage, and/or challenge different technological and scientific realities and their capacities to impel new geographies of monstrous desire and affect/effect. Though "art 'competes' and 'cooperates' only with other art practices, as science, specific scientific doctrines, techniques, and principles, 'compete' and 'cooperate' only with each other" (GROSZ 2008, 26), we, as geographers – as a particular assemblage of artists, scientists, and philosophers – must literally battle science in its limiting, manipulative and hegemonic forms over economic, spatial and social influence in (if not control over) the changes and transformations of our cities and our world(s).

REFERENCES

AITKEN, S. C. (2006): Leading men to violence and creating spaces for their emotions. *Gender, Place and Culture* 13(5), 491–507.

BAUDRILLARD, J. (1983): *The Transparency of Evil: Essays on Extreme Phenomena.* London and New York.

BUKATMAN, S. (1990): Who Programs You? The Science Fiction of the Spectacle. KUHN, A. (Ed.): *Alien Zone: Cultural Theory and Contemporary Science Fiction Cinema* pp. 196–213. London.

BOGUE, R. (2003): *Deleuze on Cinema.* New York.

BRUNO, G. (2002): *Atlas of Emotion: Journeys in Art, Architecture, and Film.* London and New York.

CRONENBERG, D. (1983): *Videodrome: The Criterion Collection.* Criterion. 2004. DVD.

CURTI, G. H. (2008a): The ghost in the city and a landscape of life: a reading of difference in Shirow and Oshii's Ghost in the Shell. *Environment and Planning D: Society and Space* 26(1): 87–106.

CURTI, G. H. (2008b): From a Wall of Bodies to a Body of Walls: politics of affect | politics of memory | politics of war. *Emotion, Space and Society* 1(2): 106–118.

DEBORD, G. (1983): *The Society of the Spectacle.* London.

DELEUZE, G. (1989a): *Masochism: Coldness and Cruelty.* New York.

DELEUZE, G. (1989b): *Cinema 2: The Time-Image.* Minneapolis.

DELEUZE, G. (1995): *Negotiations 1972–1990.* Columbia University Press, New York.

DELEUZE, G. (2003): *Francis Bacon: The Logic of Sensation.* Minneapolis.

DOEL, M. (1999): *Postructuralist Geographies: The Diabolical Art of Spatial Science.* New York.

GROSZ, E. (2008): *Chaos, Territory, Art: Deleuze and the framing of the earth.* New York.

HARAWAY, D. (1997): Modest_Witness@Second_Millennium.FemaleMan_Meets_OncoMouse: Feminism and Technoscience. London.

HUBBARD, P. (2006): *City: Key Ideas in Geography.* New York.

KRISTEVA, J. and L. ROUDIEZ (1982): *Powers of Horror: An Essay in Abjection.* New York.

LEFEBVRE, H. (1996): *Writings on Cities.* London.

MASSEY, D. (2005): *For Space.* London.

MASSUMI, B. (1992): *A User's Guide to Capitalism and Schizophrenia: Deviations from Deleuze and Guattari.* Cambridge, MA.

MITCHELL, D. (2000): *Cultural Geography: A Critical Geography.* New York.

SHIELS, M. (2009): 25 September 2009. Future is TV-shaped, says Intel. *BBC News*, http://news.bbc.co.uk/2/hi/technology/8272003.stm (accessed September 25, 2009).

SILVERMAN, K. (1988): *The acoustic mirror: the female voice in psychoanalysis and cinema.* Bloomington.

WHATMORE, S. (2002): *Hybrid Geographies.* London.

CONTRIBUTORS

Stuart C. Aitken is Professor of Geography at San Diego State University. His books include *Border Spaces and Revolutionary Imaginations* (with Fernando J. Bosco, Kate Swanson and Tom Herman, Routledge 2011), *The Awkward Spaces of Fathering (Ashgate, 2009). Global Childhoods* (with Lund and Kjorholt, Routledge 2008), *Philosophies, People, Places and Practices* (with Gill Valentine, Sage 2004), *Geographies of Young People: The Morally Contested Spaces of Identity* (Routledge 2001), *Family Fantasies and Community Space* (Rutgers University Press, 1998), *Place, Space, Situation and Spectacle: A Geography of Film* (with Leo Zonn, Rowman and Littlefield, 1994) and *Putting Children in Their Place* (1994, Washington DC: Association of American Geographers). He has also published widely in academic journals including the *Annals of the AAG, Geographical Review, Antipode, The Professional Geographer, Transactions of the IBG, CaGIS, Society and Space, The Journal of Geography* and *Environment and Planning A* as well as various edited book collections. His interests include film and media, critical social theory, qualitative methods, public participation GIS, children, families and communities. Stuart is past co-editor of *The Professional Geographer* and current North American editor of *Children's Geographies.*

Michaela Benson is a lecturer in Sociology at the University of York, UK specializing in the study of class and space. She previously worked at the University of Bristol and held the prestigious *Sociological Review* Fellowship. A leading scholar in the field of lifestyle migration, she is the author of *The British in Rural France* (2011, Manchester University Press), an ethnography of the everyday lives of British middle-class migrants living in southwest France, that was shortlisted for the 2012 British Sociological Association Philip Abrams Memorial Prize, the co-editor of *Lifestyle Migration* (2009, Ashgate), and several journal articles on this social phenomenon. She has conducted extensive ethnographic fieldwork with lifestyle migrants in both France and Panama, and continues to write on these populations. She is currently working on the relationship between the middle classes and urban space through her involvement in the project 'The Middle Classes and the City: Social Mix or People Just Like Us'.

Fernando J. Bosco is associate professor of Geography and co-chair of the interdisciplinary major in urban studies at San Diego State University. He obtained both his M.A. and Ph.D. degrees at The Ohio State University, and his B.A. at Wittenberg University. His research focuses on the geographic dimensions of collective action, the relations between place and the politics of memory in urban settings, and the social and political geographies of children, families and communities. His articles have been published in academic journals such as *The Annals of the Association of American Geographers*, *Social and Cultural Geography*, *Global Networks, Gender, Place and Culture* and *Children's Geographies*. Professor Bosco was born and grew up in Argentina, and regularly travels there to conduct fieldwork and research. One of his current projects involves an analysis of the relations between human rights activism, the politics of remembering and the creation of memorials and places of memory in Buenos Aires.

Jim Craine is an associate professor at California State University, Northridge. His specializations include cultural geography, media geography and the geographies of Oceania and Africa. He is currently a co-editor of *Aether: The Journal of Media Geography* and the editor of *The Yearbook of the Association of Pacific Coast Geographers*.

Giorgio Hadi Curti received his Ph.D. from the San Diego State University – University of California, Santa Barbara Joint Doctoral Program in Geography in 2010. As a social/cultural and media geographer working through poststructuralist sensibilities, his research intersects and develops connections between contemporary feminist, Marxian and Spinozan-Deleuzian discussions on politics and embodiment with nonrepresentational approaches to landscape and media to better understand relations between material productions of culture, space and symbolic economies, embodiment, memory, affect, and emotion, and the virtual/actual. He is currently working as an Ethnographer and Cultural Geographer for HDR Inc., where he is putting to practical use his research and interests in emotional and affectual geographies and memory, identity and the politics of place. He also serves as the Review Editor for *Aether: The Journal of Media Geography* and as a Lecturer and Adjunct Professor in the San Diego State University Department of Geography.

Jason Dittmer is a Reader in Human Geography at UCL. His books include *Popular Culture, Geopolitics, and Identity* (Rowman and Littlefield, 2010), *Superpowers: Captain America and the Nationalist Superhero* (Temple University Press, forthcoming), and *Mapping the End Times: American Evangelical Geopolitics and Apocalyptic Visions* (with Tristan Sturm - Ashgate, 2010)

Colin Gardner is professor of Critical Theory and Integrative Studies at the University of California, Santa Barbara, where he teaches in the departments of Art, Film & Media Studies, Comparative Literature, and the History of Art and Architecture. He is the author of two books in Manchester University Press's "British Film Makers" series: *Joseph Losey* (2004) and *Karel Reisz* (2006), and has just completed *Beckett, Deleuze and the Televisual Event: Peephole Art* for Palgrave Macmillan (2012).

Rachel Goffe is a PhD student in Geography at the Graduate Center of the City University of New York. Her research focuses on the partisan and economic role of the informal occupation of land, particularly in Jamaica, where she grew up. She was inspired by her experience as a practicing architect to understand the processes through which space becomes site, becoming engaged with antigentrification organizing in Philadelphia before deciding on doctoral study. She has been a member of the Media Mobilizing Project since 2007.

Ryan J. Goode is currently enrolled in the San Diego State University – University of California, Santa Barbara Joint Doctoral Program in Geography. His dissertation research focuses on the relationships between place-identity formation, tourism, and resistance to displacement in Rio de Janeiro's favelas.

Steven M. Graves grew up in the 1970s watching all three channels on TV in a small factory town in Ohio. He holds degrees in Political Science, History and Geography from several large Midwestern Universities. Recently, he has been teaching economic, cultural and methods courses in the Geography Department at California State University, Northridge. His research interests are eclectic, but have been loosely focused around questions of social justice and popular culture. Most of his recent publications have addressed the spatial dynamics of short-term high cost lenders. Several of those efforts have garnered national recognition for their effect on public policy. He has also written extensively about the spatial aspects of the popular music industry, including a recent essay on the characteristics of old school hip hop. His personal ambition is to drive every back road in the United States.

Amanda Huron is a founding and active member of Radio CPR. She holds a Master's degree in urban and regional planning from the University of North Carolina, Chapel Hill and a Ph.D. in earth and environmental sciences/geography from the Graduate Center of the City University of New York. She is an assistant professor of interdisciplinary social science at the University of the District of Columbia in Washington, D.C.

Edward Jackiewicz is Professor of Geography at California State University, Northridge. His research interests include development, post-development, tourism, and lifestyle migration. He recently completed the second edition of his co-edited book (with Fernando J. Bosco) *Placing Latin America: Contemporary Themes in Human Geography* (Rowman and Littlefield).

Paul S. B. Jackson is a Canadian-funded SSHRC Postdoctoral Fellow, hosted by the University of Minnesota's Geography Department. He recently defended his dissertation at the University of Toronto, entitled "Cholera and Crisis: State Health and the Geography of Future Epidemics". His dissertation inhabits a space between health, political ecology, and urban geography, but also incorporates cultural studies, history of science, and political economy. Paul has found a historical approach indispensable for a nuanced analysis of the interplay between highly infectious disease and geography, to challenge the culture and politics of today's impending crises. His postdoctoral research project examines how health experts envisioned the pilgrimage to Mecca and migrant populations as the source of potential epidemic crises. His ongoing research program seeks to explore, in different forms and at various historical periods, how social relations are configured in the intersection between urban health and the politics of crisis (thus far, sprawl, SARS, cholera, and the "disadvantaged child" of the 1960s).

Pascale Joassart-Marcelli is Associate Professor of Geography at San Diego State University, where she conducts research on immigrant integration, informal work, and low-wage labor markets. Her work focuses on understanding how place and space shape exclusion from decent jobs, healthy neighborhoods and social resources, especially for women. She is particularly interested in the role of social networks, nonprofits and community initiatives in enabling inclusion or (re)producing exclusion. She recently co-edited a volume on informal work and has published over 25 book chapters and articles in peer-reviewed journals, including the *Annals of the Association of American Geographers*, *Urban Geography*, *Feminist Economics*, *Space and Polity*, *Environment and Planning A* and *Health and Place*. She is currently working a research project on food, identity and place, which is funded by the National Science Foundation.

Brenda Kayzar is an Assistant Professor in the Department of Geography at University of Minnesota. She is an urban geographer and teaches undergraduate students the department's Urban Studies Program, as well as teaching and advising graduate students in Geography. Her book, *The New Era of Mixed Uses, 1986-Present*, which is included in the four volume set, *Homes through American History*, examines the rise of multiuse buildings and mixed-use landscapes in revitalizing older neighborhoods, redeveloping downtowns, and suburban green spaces. Her research has focused primarily on revitalization with an aim to address policy and developer practice impacts to lower-income populations, housing affordability, and landscape outcomes. Currently, she is researching the interplay between revitalization and the environmental justice movement.

Amy Siciliano holds a Ph.D in Urban Geography from the University of Toronto. She currently directs the City of Thunder Bay's Crime Prevention Council and is a Research Fellow at Lakehead University's Advanced Institute of Globalization and Culture.

Kate Swanson is an assistant professor of geography at San Diego State University. Her research focuses on issues surrounding exclusion and marginality, particularly in terms of migration, indigenous peoples, and youth. Some of her publications have appeared in *Antipode, Gender, Place & Culture, Geoforum, Children's Geographies* and *Area*. She is author of *Begging as a Path to Progress: Indigenous Women and Children and the Struggle for Ecuador's Urban*

Spaces (2010) and co-editor of *Young People, Border Spaces and Revolutionary Imaginations* (2011).

Todd Wolfson is an Assistant Professor in the Department of Journalism and Media Studies at Rutgers University. His research focuses on the convergence of new media and social movements and he is currently finishing a manuscript called *Cyber Left: Indymedia and the Making of 21st Century Struggle.* The Social Science Research Council and the National Telecommunication and Information Administration have supported Todd's research. Todd is also a co-founder of the Media Mobilizing Project (www.mediamobilizingproject.org and www.media mobilizing.org), which uses media and communications as a core strategy for building a movement to end poverty led by poor and working people in Philadelphia and across the region. MMP has been recognized as a national leader both in using media as an organizing tool and in advocating around the intersection of poverty and technology.

Place, Television, and the Real Orange County

By Ann Fletchall, Chris Lukinbeal
and Kevin McHugh

Media Geography at Mainz – Volume 2

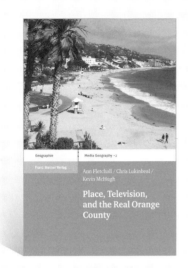

Franz Steiner Verlag

Place, Television, and the Real
Orange County
By Ann Fletchall, Chris Lukinbeal
and Kevin McHugh

2012.
144 pages with 84 illustrations.
Softcover.
ISBN 978-3-515-10118-9

This case study is an exploration of televisual place, of Orange County, California, and three popular U.S. shows set therein: *The OC, Laguna Beach: The Real Orange County*, and *The Real Housewives of Orange County*. Place is a meaningful experience in the world, and it is made through a unique intersection of social processes. It is also an amalgam of memory, emotion, and imagination. Places of the media fit this description.

Mediated places are an inextricable part of our daily lives, and directly engage the processes of place-making by affecting our perception, senses, and subjectivity. These three Orange County based series are used to demonstrate how production techniques contribute to the place-making process and how this process continues and culminates with audience engagement. The use of landscape images, the concept of emotional realism, and reality TV's claim to the real are explored for their role in the televisual place-making process. Audience surveys and the phenomenon of TV-induced tourism demonstrate the importance of televisual places to viewers. This book proves that mediated places matter.

....

Contents

....

Franz Steiner Verlag
Birkenwaldstr. 44 · D – 70191 Stuttgart
Telefon: 0711 / 2582 – 0 · Fax: 0711 / 2582 – 390
E-Mail: service@steiner-verlag.de
Internet: www.steiner-verlag.de